The Sinclair Saga

Mark Finnan

Formac Publishing Company Limited
Halifax, 1999

ILLUSTRATION CREDITS

All photos by author with the following exceptions: N.S. Department of Natural Resources: p.9; Bruce Keddy: p.14; Joan Harris: pp. 19, 21; Elizabeth Ross: p.29; Bill Sinclair: pp.39, 49, 82, 83; National Library of Scotland: p.40; Rare Books Division, New York Public Library: p.54; Tim Daly: p.66; John Richie: p.84; Gertrude Johnson: p.87; Malcolm Pearson: p.96; Dale Williamson: p. 139; Ron Norquay & Scott Sinclair: p.143

Formac Publishing Company acknowledges the support of the Department of Canadian Heritage, Canada Council, and the Nova Scotia Department of Education and Culture in the development of writing and publishing in Canada.

Canadä

Cover illustrator: Paul Tuttle

Dedicated to questors past and present

Canadian Cataloguing in Publication

Finnan, Mark

The Sinclair saga

Includes bibliographical references.

ISBN 088780-466-7

1. Sinclair, Henry, Sir, 1345-ca.1400. 2. America — Discovery and exploration — Italian. 3. Explorers — Scotland — Biography. I. Title.

E109.I8F56 1999 970.01'1 C99-950131-3

Published by
Formac Publishing Company Limited
5502 Atlantic Street
Halifax NS B3H 1G4

First published in the United States in 2000
Distributed in the United States by
Seven Hills Book Distributors
1531 Tremont Street
Cincinnati, Ohio, 45214

Contents

Acknowledgements

Innumerable personal contacts were made prior to and during the writing of this book with people I had never met before and who generously gave of their time, knowledge and resources in the process. I would like to acknowledge the following in this regard and regret any omissions I may have made. Atilla Arpat once again shared with me his in depth knowledge of Masonic architecture. Donald Bird of Truro willingly loaned me his papers on Pohl's research in Nova Scotia and other related matters. Bill Sinclair kept me well informed of the planned international celebrations for 1998 in Nova Scotia and also suggested that I participate in the Sinclair Symposium in Orkney. Niven Sinclair invited me and was a most generous host. D'Elayne Coleman was always ready to bring me up to date on the activities of the Prince Henry Sinclair Society and Pete Cummings kept me informed about happenings in the US.

Elizabeth Green of Claygate, Surrey, added a sister's care to my stay in the London region. Rory Sinclair's home was an island of overnight conviviality and comfort at the end of a hectic Toronto day. Scott Sinclair was ever obliging on a number of matters. Elizabeth Ross of Halifax shared not only her memories of her visit to Rosslyn but also her collection of slides. Joan Harris was highly informative about matters relating to her discoveries at New Ross. Lloyd Dickie, Bruce Keddy and Gerald Keddy were forthcoming more than once with details of their dig on the site. Dale Williamson of New Ross was always responsive to my queries about New Ross matters as was local historian Ron Barkhouse. Alva Pye tolerated my unannounced visits to his New Ross home. Barbara Holzmark provided me with useful information about Glastonbury.

Judith Fisken of the Friends of Rosslyn was a pleasure to follow and listen to as she brought the carvings in Rosslyn Chapel to life and Stuart Beattie, Project Director with the Rosslyn Chapel Trust, took time during a busy day to bring me up to date on the work in progress on the chapel. Josh Gourlay of Kirkwall, Orkney, was a helpful local adviser. Professor Peter Waddell of the University of Strathclyde, Glasgow, readily shared with me his knowledge of Baltic politics and pirates in Henry Sinclair's time. Robert Brydon the Scottish Templar Archivist enthusiastically took me on a highly informative tour of the preceptory ruins at Temple (Ballantrodock).

Tim Wallace Murphy, Templar author and historian, speedily responded to my many queries on Templar matters. For information on the current activities of Knights Templar within the Masonic Order in Nova Scotia I am indebted to Bob Northup, Grand Secretary of the Grand Lodge, Halifax and to Lawren Armstrong and Lindsay Woodhams also of Halifax. Brock H. Dickinson of Digby clarified for me the origin and function of the Sovereign Military Order of the Temple of Jerusalem, to which he belongs.

Dr. Peter Anderson, the Deputy Keeper at the Scottish Record Office in Edinburgh, provided me with his lecture notes from the 1997 Sinclair Symposium in Orkney which were highly informative of Earl Henry Sinclair's castle and rule in Orkney. Brian Smith the Shetland Archivist, who argued against the transatlantic voyage, extended me the same courtesy.

Tony Campbell , Map Librarian at the the British Library, not only listened patiently to my requests but personally selected several volumes for me to explore. Peggy Ann Neil of Norfolk gave me space in her home, found me the Mariners' Museum in Newport News and also got me to the airport on time for my flight back to Halifax. Donald Colp made my first visit to Guysborough Harbour an instructive one, made all the more enjoyable by the great Sunday brunch prepared by his wife Jenelle. Glen Penoyer generously permitted me to view and photograph his excavations at New Ross, was forthright about his findings and was graciously receptive to a few late suggestions. Ron Norquay willingly shared the details of his discovery of the etching of the Sinclair ship in an Edinburgh pub.

I enjoyed the company of Madeline McGowan and Marie Sinclair during our overcast, rain-swept Orkney sojourn. Geraldine McDowell, librarian at the A.R.E., responded keenly to my interest in the symbology of Rosslyn Chapel while staying at her Virginia Beach home and followed up by mailing me relevant material. Petra Mudie gave me the use of her Halifax apartment on more than one occasion. Anna Keefe loaned me literature on the Holy Grail. Helene Thibert understood my particular interest in this story and was wholeheartedly and patiently supportive. I greatly appreciated the interest expressed by family members in the progress of the book. Finally I am indebted to editor Elizabeth Eve for her extensive work on the original manuscript.

1

A Return to the Past

While living on the South Shore of Nova Scotia, between Chester and Mahone Bay, writing a book about the Oak Island mystery, I had been in the habit of dropping into the nearby Oak Island Inn at least once a week for a relaxing swim and sauna. Long hours of writing and a less than perfect posture had resulted in intermittent backache. Therapeutic massage from a local practitioner and exercise in the hotel pool helped. I also enjoyed the break from my reclusive routine, by having a chat with the staff and patrons. In addition to weekenders enjoying a few days by the ocean, there were occasional off-season guests drawn to the area by the allure of the Oak Island treasure. Whether they came from far or near, they were invariably annoyed at not being allowed onto the privately-owned island, which was within walking distance of the inn, there being a short causeway connecting it to the mainland. Occasionally I was able to lessen their disappointment by telling them what I knew about the geography of the island and the history of the ongoing treasure hunt.

After one such fireside chat one of the staff, a night receptionist, surprised me by asking if I thought the centuries-old mystery might have involved the medieval military order of the Knights Templar. My research at that point had excluded such a possibility. However, I knew that the Templars had occupied Jerusalem and other sacred sites of early Christendom during the era of the Crusades and were believed to have acquired treasures in the Middle East. In addition, they were known to have been linked to and influenced the formation of Freemasonry. Since I was gradually coming to the conclusion that individuals associated with Freemasonry, or some other secret organisation had been involved, to some degree, in the earthworks on Oak Island. I had to admit that there was the remote possibility of an indirect connection.

One day, while I was browsing through the local newspaper following my swim and sweat, he asked me if I knew that stone foundations of a very old building had been discovered some years earlier in a village about 30 kilometres inland, close to the headwaters of the nearby Gold River. He told me that speculation about the original inhabitants of the site ranged from Vikings of around 1000 AD to loyal followers of the Stuart kings, James I and Charles I. This being Nova Scotia, there was also the possibility of a Scottish involvement in the form of the man known as Prince Henry Sinclair who was believed to have crossed the Atlantic in 1398, almost one hundred years before Cabot and Columbus. This same prince was believed to have had some connection to the mysterious Templar knights and to the legend of the Holy Grail.

Although too incredible to believe in its entirety, the story was also too good to ignore, so I asked some probing questions. From whom had he heard all this? Where exactly was the site located and had he seen it himself? What artifacts had been found? It turned out that his information, which had come from a publication that chronicled the discovery, was more general than specific. He was able to tell me that the discovery had been made in the early 1970s in the village of New Ross in south-central Nova Scotia, but he had no knowledge of the people who had uncovered the stone walls.

In spring 1993 I began an odyssey that would take me over the next several years, not only to the site of what might be Nova Scotia's first European settlement, but back across the Atlantic and through the pages of history to walk among the majestic ruins of Glastonbury Abbey, long associated with England's legendary King Arthur, and to stand among effigies of dead knights in a twelfth-century Templar church in the centre of London. My journey also took me to the crumbling remains of a once elegant Scottish castle, home of the Sinclairs, and to gaze on the array of sacred and secret carvings in the interior of adjacent Rosslyn Chapel. Finally, I headed out into the turbulent North Sea to visit the rain-swept islands first settled by Vikings and once ruled by Norwegian kings, the islands from which Prince Henry Sinclair was believed to have set sail for Nova Scotia.

First, I needed to visit the ruins at New Ross. I was initially interested in the possibility that the site may have been built on during the early seventeenth century, since radio-carbon dating of material found on Oak Island had suggested that much work had been carried out there at that time. It seemed perfectly reasonable to assume that the people who had been responsible for creating the underground workings on the island, which involved mining and marine engineering knowledge of the day and undoubtedly took a long time to complete, could also have built an inland stronghold, especially if gold and precious stones were involved. On the assumption that there might be a connection between the two locations I decided to check out the New Ross site as soon as the weather permitted. Of course, the possibility that the site might hold the remnants of a visit to these shores by either eleventh-century Viking seafarers or fourteenth-century Templar knights was an additional incentive.

A visit to Chester's public library, which occupies just three rooms of a compact, period cottage on a hilltop overlooking the coastal inlet, provided me with a local history of New Ross. Although written in a formal style by an elderly resident, it was packed with pertinent details of the village's beginnings in the early nineteenth century. Being a fairly slim volume, I read through it while enjoying a leisurely lunch at Chester's acclaimed Lost Parrot Café. By the time I was ready to take a morning off from my writing, and the weather seemed settled enough for me to drive northwards, I had reread it several times and gained a good deal of background knowledge on the growth of the settlement during the past two centuries.

Although tourist maps now give the name New Ross to the village, this name really applies to the entire region which stretches inland from the Chester district to the north eastern corner of Lunenburg County. The village itself, which lies roughly in the centre of this region, is named Charing Cross but is often called simply "the Cross." As I discovered during my drive into the area, it was a collection of original settlement clearances, also known as sections, scattered across a rugged, hilly landscape spotted with lakes, streams, bogs, barrens and forest. The area was first surveyed around 1775, just 26 years after the founding of Halifax and about 15 years after Chester had been settled, mostly by New Englanders. There had been plans to cut a military road through the area directly connecting the new capital on the province's eastern shore with the former one at Annapolis Royal, in the southwest. However, the road, little more than a track through the forest, was only partially completed. The first settlers did not venture into the interior for another 40 years by which time the settlements of Chester, Mahone Bay and Lunenburg were well on their way to becoming thriving coastal communities.

In the summer of 1816, to this still remote and physically challenging region some 175 former British soldiers from disbanded Nova Scotian and Newfoundland regiments came to clear the land and make settlements. It was a time when poverty, hunger and disease, as well as the vicissitudes of the Napoleonic wars, caused many to opt for the offer of free land and the hope of a better life on a new continent, far across the Atlantic.

Governor Sherbrooke and his successor Lord Dalhousie were determined to see the interior of the province gradually cleared, cultivated and populated. Most of the best agricultural land in the province, in the north west and the south east, had been divided up between New Englanders, Loyalists and German immigrants, following the founding of Halifax in 1749. Although the many fishing harbours along the south shore had been taken over by seasonal fishermen from the Boston region, the least fertile land in the interior remained uncultivated.

One of the main resources in the region was trees. During the first one hundred years of settlement, they were the backbone of a thriving shipbuilding industry. With muscle and a mill, the settlers found the forests to be a ready source of income. Many families became almost self-sufficient with a few sheep, a cow or two, a good garden and some woodland. The newly disbanded and war-weary soldiers, most of whom had no other alternative, were willing candidates. The government's offer of a parcel of land was sweetened by the promise of free rations for a year that also included a steady supply of rum.

The eastern end of the proposed Annapolis road was not cleared so the men had to walk through the thick bush from the seashore, at Chester, with whatever possessions and supplies they could carry with them. Since many of these ex-service men were city-born, they were unaccustomed to the work involved in felling trees, clearing the stumps, building homes, planting and harvesting crops and caring for livestock. The determination necessary to 'tame' the wilderness was too much for many of them. Nor did they have the skills necessary for trapping in the woods or fishing the rivers and lakes, in spite of the abundance of fish, flesh and fowl.

Governor Lawrence, an administrator of the province and a military man himself, had encountered this problem during the settlement of Halifax. Referring frustratingly to the policy of trying to turn disbanded soldiers into productive settlers as "the King's Bad Bargain," he wrote to the

Lords of Trade and Plantations in London with the following observation: "According to my ideas of the military, which I offer with all possible deference and submission, they are the least qualified, from their occupation as soldiers, of any men living to establish a new country, where they must encounter difficulties with which they are altogether unacquainted." His remarks had obviously been ignored by his successor, Lord Dalhousie, who may have been concerned that if such an offer was not made to the former British troops they might make their way to the rebel American colonies and who knows what might transpire.

Not surprisingly, given their former occupation, dispositions and the day-to-day hardships, there was no way many of the settlers could have achieved self-sufficiency within one year. It was a daunting prospect for even the hardiest pioneers, so a fifteen-month additional supply of rations was wrangled from the Halifax administration to avoid starvation. These came to an end in the autumn of 1818, the run of rum having ceased the previous July on the orders of the somewhat disillusioned governor. There was also the fact that, as was often the case in settler communities, women were in short supply. The comforts, companionship and help they could have supplied were sorely missed. In fact, records show that during the first few months only one woman, Ellen Donnellan, who hailed from New Ross in southern Ireland, braved the wilderness. In time other wives of settlers arrived, of course, but they remained few in number and it is doubtful whether any single women joined the settlement during its first few years.

As was to be expected, of the more than 100 grants issued in 1816 only 67 remained occupied three years later. A large number of the original settlers abandoned their lots or sold them for a trifle. Replacements came in due course and this struggling settlement on the south-facing slope of Nova Scotia's east-west watershed took root and grew.

The purpose of my journey was to view a site that supposedly predated the founding of this settlement. From what I already knew I surmised the 'ruins' at New Ross were at least 200 years older. Although I wondered whether the village's original name of Charing Cross had been taken from the well-known London location, I was even more curious to know how the name New Ross, an historic Irish town with which I am familiar, had found its way from Ireland. Had it been so called to honour the first woman who bravely accompanied her husband to an unknown future in the challenging interior or was it because of Captain William Ross, also Irish and from County Cork, who had been appointed to head up the settlement? The official history suggested otherwise. Apparently the settlement's first name, Sherbrooke, had been claimed by another part of Nova Scotia and a new name had to be found. It came courtesy of a provincial administrator, Lord Mulgrave, who held land in the vicinity of Ireland's New Ross.

A turn on the road brought me to a bridge crossing the rock strewn Gold River. I stopped the car for a look at the fast flowing water below. The river, with its source further to the north-west, runs a serpentine course south to enter the ocean three kilometres east of Oak Island. Although it seems too shallow for the purpose, the river had once provided a convenient run for loggers clearing the surrounding countryside. Lumber companies had set up camps in the area after the settlement had been established and log drives in spring had sent valuable heavy timber from the New Ross hillsides down to a mill at the river's mouth. The birch and maple were used for shipbuilding, the spruce and fir for barrels for the province's burgeoning fish trade. After the first stand of hardwoods had been cut, softwood followed the same course ending up as part of the province's ever-increasing export of timber.

According to local history, the river received its name from the gold found along its lower banks by early French settlers. From where and when might they have arrived here? With the explorers Sieur de Monts and Champlain, who landed on the south shore in 1604? From Count de Razilly's settlement of 1632-36 at the mouth of the La Have river a few kilometres to the south? His enterprising associate, Nicolas Denys, had sailed into Mahone Bay and explored the coastline. Perhaps he and others had found the river and seen traces of gold.

In fact, gold in profitable quantities was mined along the river during the latter half of the nineteenth century. Some small fortunes were apparently made, but the find eventually ran its course and the mining operations petered out within a few decades. However, prospecting had continued sporadically and I had even heard that there was interest in reworking the vein using more efficient, modern methods.

After a few moments contemplating whether Greenland Vikings, European knights, seventeenth-century French settlers or English courtiers might have picked their way along the river's rocky banks on their way to its source, I continued on my way. Soon I came in sight of Ross Farm Museum, which was once the settlement's most prominent homestead, being that of Captain William Ross. The museum is actually a working farm that preserves the ways of an early nineteenth-century rural settlement. As such it provides the opportunity to observe the physically demanding ways of rural life in former years and appreciate the patience, perseverance and ingenuity of some of this country's pioneer settlers. I stopped the car and watched a lad lead a hefty oxen in harness out from the large, well kept barn. Sheep and other farm animals called out from the building's dark interior. I strolled alongside the fence, enjoying the sights, smells and sounds of farm life.

The tourist information at the entrance explained that Ross had been given charge of settling this region. He had certainly chosen his own site well: it overlooked a small and, no doubt, well-stocked lake. Appointed the first justice of the peace for the area, he and his family were obviously held in high regard by the governor in Halifax who made the gift of a Broadwood piano to young Mary Ross. After being shipped to Chester, the piano was carefully carried all the way back into the hilly interior on the strong shoulders of four stalwart military men. Given the difficulties and

Overhead view of New Ross, "the Cross," showing excavation site and Gold River.

hardships involved in such an undertaking, one can only hope that they were liberally rewarded from the rum supply.

William Ross had proven to be an able leader in difficult circumstances and had obviously been successful in his own efforts to settle here. Unfortunately, his attempt to personally persuade the authorities to complete the road from Annapolis to Halifax resulted in his untimely death. In autumn 1821 he visited Halifax to remind the governor of an earlier promise to have a road built linking the settlement with the capital. His business done, he set out with an Indian guide determined to cut a line back through the thick forest. During the night a heavy rainstorm struck and the severe drenching led to pneumonia and his demise in a Halifax hospital a few months later. His wife struggled on at Rosebank, the name they had given to their homestead, along with her seven children. All but one survived and became prominent members of the new community.

Following a short drive further up the road, I finally sighted the Cross. Passing the school to my left I noted that the village was situated about halfway up on the southeast side of a hill. Its location gave a commanding view of much of the surrounding countryside and looking south one could see the waters of a nearby lake and the river.

Unsure of the exact location of the site, other than that it lay in the back of a property in the village itself, I pulled into the gas station and as nonchalantly as possible inquired about rediscovered ruins. The puzzled attendant scratched his head and looked off into the distance as if desperately trying to find a helpful answer. He finally admitted he had never heard of such a thing and suggested that I should ask someone else. I ventured beyond the crossroads and pulled up in front of a grocery store. I waited until an elderly customer came my way. After smiling gently at me, in response to my question, she pointed towards a house further along on the opposite side. I thanked her and moved on.

A few moments later I passed by the nondescript house, surprised to see a realtor's sale sign in the front yard. I drove on as far as the agricultural fairgrounds, turned around and headed back to the property. On entering the driveway, I saw that the house was empty and that the back garden, where the alleged stone foundations were discovered, rose steadily away from the rear of the building. The entire property had quite a forlorn appearance. After parking the car I hesitated to investigate without permission, but then decided to take the plunge.

Walking up through the garden area I could not see any surface signs of the remains of an earlier structure. The only object of possible archaeological significance was a waist high, upright stone which was supported at the base by several smaller ones. I had expected to see foundation walls and felt a little disappointed. Perhaps I had been directed to the wrong place or had misunderstood the directions. The thought crossed my mind that I had been conned. Had

I come here looking for something that never existed? I had to admit that my interest in the Oak Island mystery set me up as a prime candidate for just such a prank. Finally it occurred to me that perhaps what I was hoping to see had been covered up again by the former owner, or was out of sight further up the scrub-covered hillside.

Since the afternoon was getting on and it was becoming noticeably chilly, I decided to forego clearing away any surface soil or investigating beyond the bushes at the rear of the garden. After looking inquisitively through one of the windows of the house, I noted the telephone number of the real estate agent and returned to the comfort of my car. On the way back I decided to call the agent in the hope of contacting and talking with the owners about their discoveries. I even debated whether, based on what they might tell me, I should explore the possibility of renting the place. The matter was settled for me a day or two later when the agent called back to tell me that the owners, who had left the province, did not want to discuss the property's history and only wanted to sell it. I put the New Ross story on the back burner with the vague intention of pursuing it sometime in the future, should some further information happen to come my way.

Come my way it did and sooner than I expected. A visit to the Oak Island Inn a few weeks later landed the book *Holy Grail across the Atlantic* in my lap. Published just a few years earlier, it was the work of American-born Michael Bradley who had previously written about possible pre-Columbian voyages to North America. This highly speculative take on early Nova Scotia history, interwoven with earlier proposals by a trio of English authors about the nature of the Holy Grail, contained a mixture of detailed and disguised material about the New Ross site. While living in Nova Scotia in the early 1980s, Bradley had visited the site at the request of the owners, a middle-aged couple to whom he gave the pseudonym McKay. This coincidentally happened to be the name of the first known settler on the site, Daniel McKay, the settlement's first blacksmith.

Museum personnel in Halifax had dismissed the McKays' discoveries as being of native or French origin and therefore of little significance. But since the Mi'kmaq of Nova Scotia were not known to have built with stone and there were no records of French settlers in the area prior to the New Ross settlement of 1826, neither of these suggestions seemed to make sense.

Having talked with the couple and seen clear evidence of rubble stone foundations in their back garden, Bradley had come to his own conclusion as to their origin. As I was later to learn, this did not sit too well with the McKays, whose real name turned out to be Harris. However, what interested me initially in the book was not Bradley's own surmising that Prince Henry Sinclair had visited the site, nor his second-hand, sensational theory associating the Holy Grail with a blood line extending back to the early days of Christianity. The book contained a few photographs of the exposed foundations and some references to Joan Harris's own conclusions as to their possible origin. The historical time-frame and the people involved in her theory about one of three structures she claimed had once existed at the rear of their property dovetailed neatly with my own conclusions about the origins of the Oak Island mystery.

This of course led to a longer and more focused second visit to New Ross soon afterwards, during which closer examination of the site enabled me to make out half-buried rock clusters which might have once formed part of the walls of a small building. My curiosity somewhat satisfied, I left the property realising that without any direct input from the Harrises or an archaeological investigation of the site, there was nothing more I could learn about it.

I had no idea at the time that I would return to New Ross on many more occasions during the next few years and that I would be given a copy of Joan Harris's hand-written notes, outlining in her own words what she had discovered in her back garden. Nor did I expect that one day I would actually see an archaeological dig taking place on the site.

2

Stones and Stories

Preoccupied as I was at the time with researching and writing about the Oak Island mystery, I had little inclination to embark on any additional research or exploration. So as compelling as it might be, the possibility of there having been pre-Columbian visitors to Nova Scotia had to wait my serious attention until later. The notions being bantered about that Carthaginians, Celts, Vikings, Venetians and a medieval prince had found their way to these shores did not seem all that far fetched to me. In fact, I strongly felt that they merited further investigation by someone who had the time, the interest and the resources to do so.

In one sense I was relieved not to have been sidetracked from my focus on the historical timeframe scientifically established for much of the work on Oak Island and my research into the lives of the cast of sixteenth and seventeenth-century characters I had concluded were most likely responsible for hiding the mysterious treasure there. However, Joan Harris's claim that a seventeenth-century manor style house had once stood on the New Ross site held my interest. If such a claim ever turned out to be true then the site quite possibly had some connection to the people who had dug the pit on Oak Island.

Gerald Keddy, left, and Lloyd Dickie, right, excavating below the 'Standing Stone' at New Ross.

Although I was totally convinced that there was no connection whatsoever between any Sinclair voyage and the Oak Island mystery, it was becoming obvious to me that many people were interested in that possibility. And what was equally if not more observable to me was the fact that this interest was predominantly driven by the fascination that there might be a linkage, however obscure, to the heroic and romantic legend of the Grail.

As a result of public presentations I gave about the Oak Island mystery in 1994, I was approached by Gerald Keddy, a New Ross area tree farmer, Lloyd Dickie, a retired government scientist with an interest in archaeology, and Dr. Bruce Keddy, a former medical man turned gold prospector. In the course of our discussions I learned that they had carried out a private dig on the New Ross site in 1993. In the process they had been able to dismiss one of the many rumours about the site while at the same time bringing to light solid evidence in support of at least one other.

Having kept abreast of the media reports and the published account of the Harrises' discoveries, as well as the ongoing speculation about the site, they had arranged for permission to dig beneath the stumpy standing stone on the property. Supported at the base by a circle of much smaller stones, it had somehow become rumoured as marking the spot of an ancient holy well. Gerald, who had grown up in the area and had his own views about the site, joined Lloyd and Bruce in removing the upright stone and digging deep beneath it. They found no trace of a well or even of water. They dug a hole more than three feet deep and found that the soil several feet down looked as if it had never been disturbed. However, they did find ashes and pieces of charcoal in the soil below some of the circular stones.

Lloyd had the charcoal radio-carbon dated and Bruce had a sample of a fine sand found on the site tested. The result of the radio-carbon dating was a surprise to all of them. It indicated that the charcoal found among the ashes was as old as 1500 BC. The test on the sand showed that it contained an unusually high concentration of gold dust.

So what, if anything, did these results suggest other than that the site had possibly been used by native inhabitants some three and a half thousand years ago? The high concentration of gold dust suggested to Bruce Keddy that either there could have been a gold refinery of some kind on the site centuries earlier or, impossible as it may seem, that the hillside had once been part of a gold-bearing river bed. Although little or nothing had been definitively proven, at least the rumour about the stone-covered well had been finally put to rest.

In 1995 I came upon a copy of Joan Harris's detailed notes relating to the years she and her husband Ron had lived in New Ross. Barbara Holzmark, a local craftsperson with an interest in things Celtic, had obtained the pages from Joan during a visit to the site in the late 1980s. It was an uncoordinated written commentary, but it helped clarify the growing confusion about the site caused by constantly overlapping layers of speculation, most of which bore little relation to what had actually been discovered and documented. There was also a long list of blistering criticisms of Bradley and his book, Joan believing she had been badly served by him and regretting having confided in him. These notes were obviously intended for a book of her own.

The Harrises arrived in New Ross in 1972, when Ron was offered a teaching job at the local school. They had come there from the coastal town of Shelburne which lies some 160 kilometres to the south. They had spent many years globe-trotting, including stints in Japan and India where, according to Joan, they had been engaged in social work. They appeared to be an energetic couple with lively minds and a variety of interests and likely to be considered a little eccentric by the locals.

While living in Shelburne they had been in the habit of exploring the surrounding countryside, mostly by walking down long abandoned back roads and little-used forest trails. It was while they were on one of these walks that they first heard the story about a large house having been built in the woods in central Nova Scotia in earlier times. An elderly Mi'kmaq, whose lakeside hut they had accidentally come across, told them that crown land, so designated in the first Nova Scotia charter, had been intended for the sole use of the native population and the monarchy itself. According to another Mi'kmaq they met, a mansion of sorts had been built in the wilderness on high ground, west of present-day Halifax, long before the city's founding in the mid-eighteenth century. The mansion had been intended as refuge for the English king at the time. To Joan Harris who had an avid interest in the secrets and shenanigans of British history and a willingness to take such a tale quite seriously, the king in question was either the Scottish born son of the executed Mary Queen of Scots, who early in the seventeenth century became James I of England, or his son, Charles I, whose fractious relationships with Parliament, religious extremists and Oliver Cromwell led to his death by execution. Both kings had been extremely interested in and supported the efforts of their advisor and close friend, Sir William Alexander, to establish settlers in the newly named Nova Scotia. From what I knew about Alexander and his lost colony of Charlesfort, I was aware that his unique position at the Stuart court, combined with the powers given him in the Nova Scotia charter of 1621, certainly allowed for some clandestine activity on behalf of the crown.

When the Harrises first made enquiries about possible houses to purchase in New Ross the realtor had casually mentioned that the village was located on one of the highest points west of Halifax. This passing comment, so similar to one she had heard earlier, was not lost on Joan. They settled for renting a property in the centre of the village near to Ron's school and while he was negotiating the rent with the owner, Joan ventured onto the badly overgrown piece of land at the rear of the house checking out the prospects for a future garden. While wandering through the weeds she happened to notice something that held her attention. As she headed up a narrow pathway that led up to the rear of the hillside lot, she saw a line of flat, round stones embedded in the ground. Sensing that she was standing on a mound overlooking the surrounding landscape and recalling the story of the mansion built on high ground west of Halifax, she began to wonder if they had accidentally stumbled upon the site of the former mansion mentioned by their Mi'kmaq friend.

After settling in, the Harrises heard from one of Ron's Shelburne pupils, a lad named John Nauss, that something resembling a castle had once stood on the property they had rented. Nauss told them that his family, his grandfather in particular, knew of stories about the building and of its destruction. According to his grandfather's tale, an ancestor had been brought to Nova Scotia in the 1620s to help build the mansion or castle. There had even once been a drawing of the mansion in his family's possession. Unfortunately it had long since disappeared with a relative who had moved to the United States. And according to another of grandfather Nauss's stories, a later relative, this one being on his wife's side, had been involved in its demolition. The young Nauss later produced a rough sketch of the structure complete with a front balcony supported by twelve marble pillars and a gold covered dome atop the roof. According to the Nauss tale, these had disappeared with the destruction of the building, carried out on the orders of none other than Oliver Cromwell himself. Nauss claimed that these materials reappeared as part of the State House in Boston.

All this might have sounded implausible to some people but to Joan Harris it was the stuff of history, a conspiratorial history. It was Charles I who had renewed and expanded Alexander's Nova Scotia charter and supported efforts to establish the first English settlement on its shores. Charles, who was of Scottish extraction, was also inclined to be too sympathetic to the Catholic cause and had ties by marriage with England's traditional political and religious enemy, France. With his life in danger, he might well have planned a refuge in the New World. English and colonial history confirms the fact that Cromwell, who replaced the flamboyant monarchy of the Stuarts with his distinctly puritanical rule, was capable of ensuring such a refuge was destroyed.

Cromwell had willing accomplices in New England to carry out his orders. Commander Robert Sedgewick had set sail from Boston with an army of 500 men and the support of four warships sent out by Cromwell in 1654. Sedgewick's targets were the French at what had been Sir William Alexander's Charlesfort, present-day Annapolis Royal, and other settlements in the vicinity of Cape Sable and La Have. But Joan Harris reasoned that the New England force might also have been under secret orders to locate and destroy the Stuart refuge heard to have been built in the Nova Scotia wilderness.

All this certainly had an effect on her subconscious because she later claimed to have seen the headless ghost of Charles I wandering around the back garden of her New Ross home. But perhaps the more surprising, if less dramatic thing was that she shortly afterwards made discoveries in the same garden that added some substance to her theory.

Circumstances quickly progressed that made it possible for the Harrises to purchase the property. While poring over the early deeds they discovered that a wall of stones had

Stone foundations uncovered by Joan Harris on New Ross site.

existed on one end of the property since time immemorial. They wondered who could have put it there. By September of their first year they had cleared much of the rear lot and Joan had made her first discovery—two similar sized circles of stones, each surrounding a central hole. She drew the conclusion that these had once held the corner posts of the wooden frame of a building. The filled-in walls were made of loose stones gathered from the surrounding area. Somewhat encouraged, she then uncovered what she believed were the sunken outlines of the foundations of a 4.3-metre wide building in the centre of the garden area. It had an east-facing opening with an extension, a "portico" in Joan's words, extending from it. Joan also mentioned that Ron managed to expose a portion of one of the foundation walls to a depth of about one metre. The deeper he dug the larger the stones became and some even looked as if they had been shaped by hand.

Based on the stones and the stories she had heard, Joan quite matter-of-factly deduced that the architect must have been Inigo Jones, the famous English court architect and designer of the period. Adding to her utter conviction that this must have been so was the fact that her research indicated that Jones had been unexplainably absent from the London scene for nine years between 1623 and 1632. When the Harrises later opened their home as a tourist hostel named the Pig and Whistle, it was publicised as being on the site of a former castle.

Although no professional archaeological dig had been carried out on the site and there was little evidence to support her theory, I could not readily dismiss Joan Harris's conclusions about the New Ross ruins. However, I had to admit that the assortment of stones visible in the back yard certainly seemed to be as confusing as the contents of a pig's breakfast and there was an absence of factual data. Then some new discoveries led Joan to an entirely different historical scenario.

With the curiosity and determination of an historical sleuth, Joan had continued with her digging the following spring and eventually came across what seemed to be a series of ash pits towards the rear of the property and just west of the foundations earlier uncovered. Then just to the south of these and leading off from the side wall of the smaller building in the centre of her garden, she began to find buried traces of two other stone walls, 3.7 metres apart, running off onto the neighbouring property. She eventually determined that they were both about 23 metres in length. She also found some small man-made spruce clamps split, at one end, less than half a metre down in damp ground behind the house. An American visitor who happened to be an archaeologist suggested that they were centuries-old wooden clamps sometimes used in the shaping of hot metal into small objects. To Joan the metal must have been gold since the Gold River was close by. Therefore she reasoned the objects being shaped must have been either coins or ornaments. According to Joan, when these primitive clamps

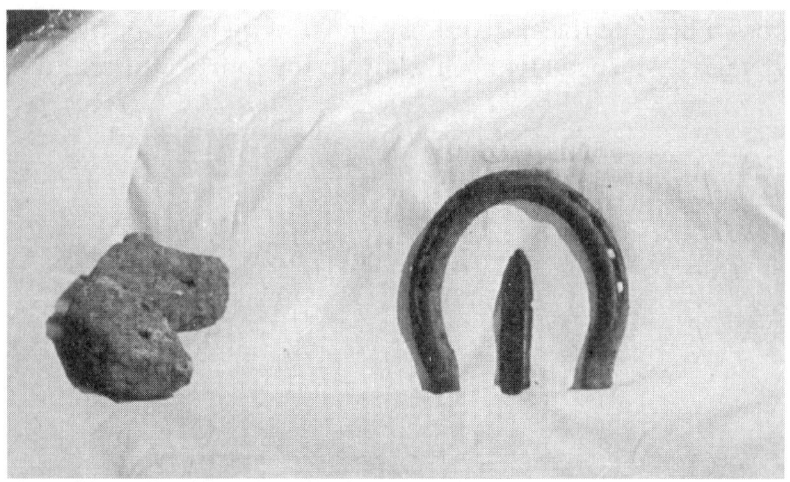

Piece of 'sword' and horsehoe found on New Ross site.

were later radio-carbon dated, they were discovered to be anywhere from 600 to a 1000 years old. In other words, as Joan was quick to point out, they could have originated with the Vikings who sailed into North American waters around 1000 AD. Her manor site appeared to have a pre-history of its own. It now took her back in time to the days of Viking voyages and to the distinct possibility that the Viking explorer Leif Erickson, or some of those who had followed him, had not only landed on the south shore of Nova Scotia but had set up home on her hillside.

The clincher for Joan was the chance discovery by a student helper of a piece of an iron sword in the ditch at the front of the house. The 13-centimetre length of blade had a depressed central channel and Joan, who claimed to have spent some time examining Viking swords in European museums, was certain that this was Norse in origin. Joan concluded that the 23-metre long building had been a Viking hall similar to those found in Greenland and in other lands occupied by the Vikings.

The unearthing of the chunky stone with the crude shape of a small cross on one end led her to believe that people of a Christian persuasion, such as the Greenland Vikings, had been there. The new evidence pointed to the possibility that the site could be linked with the transatlantic voyage by Prince Henry Sinclair, in the fourteenth century.

3

Vikings and Vinland

As I mulled over these additional discoveries and claims, I began to wonder if some clues might be found in the Viking sagas, the same sagas that could well have encouraged Prince Henry Sinclair to embark on a voyage across the North Atlantic.

My knowledge of the Vikings was mostly confined to what I had learned during my school days in Dublin, a city that actually originated as a Viking trading centre around 840 AD. Extensive Viking ruins and artifacts were unearthed in the mid-1970s in a region between Christ Church cathedral and the banks of the River Liffey, at Wood Quay, near the centre of the city. The impressive and varied discoveries there have since led to a major, authentic recreation of part of the Viking settlement, which has become one of the city's major tourist attractions. But other than some generalised and perhaps biased impressions of the Vikings as savage seafaring raiders, ruthlessly attacking and plundering coastal monasteries and villages in Ireland, England and Scotland, I was really ignorant of their considerable accomplishments as ship builders, navigators, explorers, traders, settlers, farmers, politicians, artists and poets. And more to the point, apart from a vague knowledge of the 1960 discovery of the ruins of a Viking settlement at L'Anse aux Meadows in northern Newfoundland, I knew little about the details of their many westward voyages.

While doing research for a book about the life of Sir William Alexander and the Charlesfort settlers in Nova Scotia in 1629, I had spent time in Scotland familiarising myself with its early history, which in turn gave me an insight into the Viking voyages. My flight back across the north Atlantic took me to Iceland, where the capital Reykjavik stands atop that country's first Viking settlement, then south of Greenland and on to the east coast of North America, the same route that the legendary Norse seafarers had taken a thousand years earlier. Later, on the trail of Prince Henry Sinclair, I would again follow much the same, but by then much more familiar route, aware of the men who led the Norse thrust westwards, men such as Eric the Red, Leif Ericsson and Thorfinn Karlsefni.

The L'Anse aux Meadows site on Newfoundland's northern peninsula was first discovered by Norwegians Helge Ingstad and his wife Anna Stein. Firmly believing in the veracity of the Viking sagas describing voyages to North America, Helge Ingstad headed an expedition to find the physical proof. After much time, effort and money had been spent exploring the east coast of the continent, they came to Newfoundland and with the help of a local fisherman, found the tell-tale earth covered mounds. Anna, an archaeologist, supervised the excavation, during which the remains of a settlement appeared. Norse artifacts found on the site proved beyond a doubt that the Greenlanders had been there. With this discovery the known history of western exploration was changed forever. Here was solid evidence that placed the Norse at the forefront of transatlantic exploration and as the first to bring European influences to the shores of North America. Although Ingstad and Stein remain adamant that they found Leif's final destination in North America, the legendary Vinland of Viking lore is believed by many to be much further south. This final frontier of Viking exploration was, according to the sagas, a place of natural beauty and abundance — a welcome contrast to what had been encountered elsewhere.

All of this was to stand me in good stead as I explored the material that first suggested that the Norse-appointed Earl of Orkney, Henry Sinclair, a man of Viking ancestry, had led an expedition across the North Atlantic in 1398. Having already familiarised myself with the history and geography of the Viking trail westward, the details of the much debated account of Sinclair's voyage became more plausible.

As I said, I knew all too little about how the Viking Age (750-1100) had influenced the history of Europe. I was surprised to learn that the Vikings had established trade routes that ran from the Baltic to the Middle East and that, apart from the pillaging and plundering for which they are more popularly known, they had made a unique contribution to literature and art in the Middle Ages and had established the first people's democratic parliament.

Accounts in the *Islendingabók* (Book of the Icelanders), the *Landnamabók* (Book of Settlements), *Eiriks Saga* and the *Graenlendinga Saga* (Greenlander Saga), part of a large body of Scandinavian seafaring lore, tell of adventures beginning in 700 and of the gradual shift westwards as centuries passed and ship building progressed.

In both Iceland and Greenland, where permanent settlements were established and prospered, the Norse eventually became victims of their own inability to make the adjustments necessary for survival in terrain and environments that were more demanding than those they had known in Norway. The geography, geology and limited resources of these environs presented tough challenges. Climatic change that brought a mini Ice Age eventually forced an end to their western communities. All that was left were the stone remains of settlements and, of course, the stories, the sagas, told to later generations gathered by the fire in the long halls and houses of the mother countries. These described the daring exploits of their ancestors on sea and land, in the harsh and demanding conditions that only the most hardy men and women could withstand.

According to the *Landnamabók*, the Viking voyages westward began in earnest after the Shetland and the Orkney islands, north of Scotland, and the far flung Faroe Islands had been settled, in the ninth century.

It is quite possible that the Norse may have heard of new lands to the west from tales told by Irish monks and Scottish fishermen. Perhaps they had even heard of the legendary voyage of the sixth-century Irish monk, St. Brendan. What is known for sure, from the sagas, is that Iceland was first sighted by a Norseman blown off course while on his way to the Faroes, around 860. Later a group of settlers set out from the Faroes led by a man who became known as Raven Floki. His name derived from his use of ravens to help him find land that to all appearances was uninhabited. As soon as the sea ice that filled the fjords into late spring thawed, Floki returned to Norway. There he spoke derogatively of the newly discovered land as Iceland and the name stuck. This did not deter other desperate or daring types from trying their luck in a land that actually lay on the same latitude as central Norway and was only a week's sail away in decent weather. The coastline had already been charted and it was known that there was an abundant supply of wildlife and fish. There was pasture aplenty and land covered with stands of birch and willow, much needed for fuel and building purposes. With Norwegian coastal real estate becoming scarce, it was not surprising that by 930 most of the 100,000 square kilometres of inhabitable land in Iceland had been taken over.

The first major settlement was established on the site of present-day Reykjavik and it was in this region of the partially glacier covered, volcanic and earthquake-prone island that Europe's first parliament, the Althing, evolved. A measure of the nature of Icelandic society at the time can be gleaned from a remark made by a German cleric, Adam of Bremen, who said that in Iceland the law rather than a king ruled. The concept of a democratic republic had materialised, some say out of necessity, on a remote landmass in the North Atlantic.

The hardy Icelandic settlers not only survived the experience of establishing themselves in a challenging land but they reached a measure of cohesiveness and direction as a people which, combined with a growing prosperity and political stability, evolved into that nebulous entity, more than the sum total of inhabitants, land and resources, known as nationhood.

The Icelanders exported such commodities as wool, cheese and fish and in return, imported many items from all over the known world including the Middle East. They even had a bishopric that received its authority from Rome.

Unfortunately the prosperity could not be maintained. Overcutting of the fragile forests, overgrazing the thin layer of topsoil led to an environmental crisis. Then a devastating famine depleted the population and those remaining began to look westwards wondering if they and their children might stand a better chance of survival in regions said to exist beyond the western horizon.

As so often happens in the course of human history, an individual appeared on the scene who was destined to turn hope into hard reality by leading them westwards again to an equally harsh land to which he would cleverly give the enticing name of Greenland.

Eric the Red, so called because he sported fiery red hair and a temperament to match, had been expelled from his native Norway because of his compulsion to rage at and kill those who were foolish and unfortunate enough to cross him. After landing in Iceland the chastened but astute Eric married into a wealthy family and acquired a sizeable homestead. He seemed destined for a reformed life in the upper echelons of Icelandic society. However, his Viking temper got the better of him yet again and after murdering a neighbour in an argument over borrowed timbers he was exiled once more, this time for a period of three years. But where was a man in his predicament to go? Certainly not back to Norway where, according to the law of the day, he could have been killed on the spot. Having heard stories of the accidental sighting of yet another seemingly uninhabited land to the west, he persuaded some other Icelanders to join him on a voyage of exploration. So, in 982, after several days sailing into unknown waters, they landed safely on the coast of Greenland. They established themselves in what became known as Eastern Settlement, although it actually lay on Greenland's southwest coast.

By 985 some 500 settlers had arrived and Eric, the former outlawed criminal, had now risen to the rank and status of chief, a veritable king in his own kingdom. He established his farm on prime land at Brattahlid overlooking the picturesque shoreline of Eric's Fjord. Over the next 15 years or so several other groups of settlers arrived so that another area some 500 kilometres up the coast was set up with about 90 families. Not long afterwards there were about 3000 Norse in Greenland who survived by raising sheep and small cattle, fishing and even a small amount of grain production. They were determined and seemed destined to stay. But it was not to be.

About the same time that Eric and his people were establishing themselves in Greenland, another Viking was accidentally sighting the coastline of yet another landmass even further to the west. Bjarne Herjulfsson had been on a trading mission from Iceland to Norway. On returning home he learned that his father had left for supposedly greener pastures with Eric the Red. So Bjarne followed in his wake in the summer of 985, and due to a storm that took his boat way off course, became the first European to sight the shores of North America. Bjarne, who seemed short on adventuring spirit, rejected the pleas of his men to go ashore on this wooded, hilly land which he knew could not be Greenland. Instead, he turned the ship northwards in the hope of finding his way back. Two days later he sighted land that was flat and tree covered. After three more days they came in sight of yet more land, barren and mountainous, which they also bypassed. Then after getting caught up in a gale for four days they managed to cross back to their

Replica of Viking ship which sailed from Norway to Nova Scotia.

original destination on the west coast of Greenland. Word of the new land sightings quickly spread and so was sown the seed that in the mind of Eric's son Leif would one day become a full blown passion to find and settle the tree-laden land to the southwest.

About five years after Bjarne's unplanned adventure, Leif, who had already earned his sea legs in several trading trips to Norway took it upon himself to continue the Viking tradition of discovery. As I looked at the romanticised, 10-foot bronze statue of Leif and the model of the graceful ship he used outside the Age of Exploration Gallery at the Newport News Mariners' Museum, I could easily believe that he was destined to take the final step in bridging the ocean between the old world and the new. Having purchased Bjarne's boat which, according to Viking belief, still held the memory of its former voyage and confident that a new land could be found, provided wave and wind cooperated, he had little trouble in getting a complement of 35 men to accompany him. By carefully retracing the route which Bjarne had travelled back to Greenland following his sightings, they succeeded in finding the landmass of which he had spoken.

The first one was rocky and barren and after landing on it they named it Helluland (Slab Land), which many people agree describes the appearance of Baffin Island. The second country they came to as they sailed southwards they named Markland, because it was flat in appearance and well wooded. This is considered to have been Labrador but it might also have applied to other landfalls further south where the geography matches that description. We are then told that after leaving Markland they sailed for two days before a north-east wind and sighted land with an island lying to the north of it. Could this land have been Nova Scotia or New England? To date no one knows for sure, but the *Greenlander Saga*, in giving us a detailed account of what transpired, suggests it had to be somewhere south of Newfoundland.

> They went ashore and looked about them. The weather was fine. There was dew on the grass and the first thing they did was to get some of it on their hands and put it to their lips, and to them it seemed like the sweetest thing they had ever tasted...There was no lack of salmon in the river or lake, bigger salmon than they had ever seen. The country seemed to them so kind that no winter fodder would be needed for livestock. There was never frost all winter and the grass hardly withered at all.

According to the saga, Leif and his men not only spent the winter there but they also built a number of houses. Expeditions were sent inland. On one such journey some of the men came across vines producing wild grapes. Dried samples of these grapes were brought back to Greenland the following spring, along with a cargo of timber. *Eiriks Saga* confirms this discovery and tells us: "Leif set sail when he was ready. He ran into prolonged difficulties at sea, and finally came upon lands whose existence he had never suspected. There were fields of wild wheat growing there, and vines, and among the trees there were maples. They took some samples of all these things."

We are led to believe that Leif was inclined or obliged to take care of things on the home front and did not set sail for Vinland again. However, it is quite possible that years after his return not all the voyage lore was remembered or written down. There certainly are indications of overlapping in some of the sagas, leading one to conclude that some condensing took place. We are then told that another of Leif's brothers, Thorwald, took his place and with a group of Greenlanders made a successful crossing to Vinland. Pleased with their new surroundings, they stayed for two years before Thorwald was fatally wounded by an arrow in an Indian attack. It was apparently in retaliation for the earlier killing of several natives, whom the Norse referred to as Skraelings, or wretched ones, the same name they applied to the Inuit in Greenland. After Thorwald was buried somewhere on the northeast coast, his companions went back to Greenland. We are told that the next attempt to find this fertile land was undertaken by Thorstein, another of Eric's brothers, but his ship was carried off course by storms and he had to return to Greenland.

After Thorstein's untimely death, which as far as we know had nothing to do with the failed voyage, Thorfinn Karlsefni, an Icelandic trader who was wintering over at Brattahlid in Greenland, took up the challenge. He fell in love with Thorstein's widow Gudrid, a woman reputed to be of exceptional beauty and stature and who "was very intelligent and knew well how to conduct herself among strangers." Gudrid married Thorfinn and joined Thjodhild, the independent-minded wife of Eric the Red, as one of the two most notable women in the Norse sagas. It was Thjodhild who, in spite of Eric's objections, had the first official church built in Greenland and Gudrid, her new daughter-in-law, made her mark in history as the mother of Snorri, the first child of European parentage known to have been born in North America.

Thorfinn struck a deal with Leif who leased him the houses he and his men had built earlier in Vinland. Some 60 men and 5 women, including Gudrid, set sail with livestock and supplies around the year 1010 AD. They made a successful sea crossing and as one saga briefly tells, they found Leif's houses and lived in Vinland for some three years. Another more detailed saga suggests that they first wintered over at a bitterly cold location before sailing on south to Vinland, where their child was born. The northern location referred to may well have been the settlement in northern Newfoundland.

Although there are indications that contact with the native Indians was friendly at first and that mutually beneficial trading took place, matters later deteriorated. Hostilities ensued and although the Norse were initially able to fend off attacks, they were hopelessly outnumbered in terrain that was not familiar to them. Knowing that their time was quickly running out, Thorfinn eventually decided to take the party back to Greenland. With a prized cargo of mixed goods he reluctantly returned home with the Vinland settlers.

The eleventh and twelfth centuries were the most prosperous and active for the Greenlanders. Although the natural environment was noticeably less accommodating than that in Iceland, it was possible to survive there. Apart from cattle and sheep, raised in great numbers, there was also an abundant supply of caribou, reindeer, whales and seals. Some of the men trapped for walrus ivory, furs and hides, as well as the falcons and polar bears which were prized items in the courts of Europe and in the Middle East. The houses were built of stone with sod roofs supported by driftwood beams, wood being far from plentiful in Greenland. The cattle were wintered in barns that had stone dividing walls, not all that dissimilar from the remnants of the walls Joan Harris found on her New Ross property.

Much to Eric's chagrin, Christianity took hold in Greenland with his wife's help and replaced the Norse pantheon of gods and goddesses. Many small churches were built and also a monastery and even a nunnery. In 1126 a cathedral was built at Gardar, not far from Eric's Fjord. Dedicated to St. Nicholas, the patron saint of seafarers, the fallen and picked through remains can still be seen. Even though Greenland was politically, like Iceland, a people's republic, the church became the dominant influence and the largest landowner, with the bishop enjoying the role of near monarch. In 1261, Greenland came under the rule of the King of Norway, who promised to keep the already diminishing trade route going. In return for this favour he received tax revenues and a lock on the remaining trade.

The destruction of the fragile ecosystem that had earlier occurred in Iceland also happened in Greenland. The small trees quickly disappeared, cut down for firewood and building. Constant over-grazing prevented the lean pasture lands from replenishing themselves. Other factors, such as trade competition from an alliance of German trading cities, as well as disease and earthquakes, contributed to uncertainty, economic stagnation and a reduced population.

With the arrival of the mini Ice Age, what had been at best a marginally habitable country became impossible to live in unless one was to adopt to the lifestyle of the Inuit, which apparently the Norse in Greenland never did. Consequently by 1350 the more northerly Western Settlement had been abandoned. Ivar Bardssen, a cleric, who sailed up from the Eastern Settlement that year said there was no trace of anyone. All that remained were a few wild cattle wandering through the pastures. But where did the 1000 or so inhabitants of this once thriving settlement go? The fact that no human remains have ever been found during archaeological excavations in the 1970s and the discovery in the 1990s of a wooden loom still holding an unfinished length of cloth and a few other small household artifacts suggest that the settlers may have quickly departed or been captured by the Inuit or pirates. The question lingers as to whether some of them might have ventured across the Davis Strait in a last desperate attempt to settle in the greener pastures of Vinland. A papal letter written in 1492 suggests this. Lamenting the fact that the Greenlanders had abandoned their faith, it added that they had joined themselves with the native people of America. But no evidence has come to light to confirm this.

By the end of the fourtheenth century voyages to Greenland had almost ceased and, as I learned later, it was Henry Sinclair who sent one of the last ships to its shores. Apart from accidental contact with a small community on Hvalsey, an island near the Eastern Settlement, around 1400, there are no records of remaining Greenlanders. No survivors ever returned to Iceland or Norway to tell the tale of what had happened. The fate of those who had tried to survive or flee from increasingly inhospitable conditions remains unknown to this day, in spite of the work of many archaeologists and researchers.

What we do know for sure from the Viking sagas, historical records and subsequent archaeological discoveries is that Eric the Red, Leif Ericsson, Thorfinn Karlsefni and many others succeeded in making a landfall and settling temporarily in North America. The Viking sagas do not tell us the whole story. More may come to light from documents yet to be found in Scandinavian archives, from the oral history of the Mi'kmaq of Nova Scotia and in Indian legends elsewhere on North America's eastern seaboard, as well as from the evidence already found not only in Newfoundland but also in Nova Scotia. Late in the last century, an old Norse style axe with distinctive runic markings on the blade was discovered by a farmer working a hillside field near Cole Harbour in Tor Bay, Guysborough County. No excavation has ever been undertaken in the area. Other impressive finds indicative of a Viking presence in North America, but yet to be officially accepted as authentic, have also been made in Quebec, Ontario, Maine, Rhode Island and even Minnesota.

My research made it abundantly clear that an historic precedent in transatlantic exploration had been set that could all the more easily be followed some time later by a well-positioned seafarer for whom providence would provide the incentive and opportunity. As I was about to discover, Prince Henry Sinclair was just such a man.

To my surprise, while viewing the Viking exhibition in the Mariners' Museum in Newport News, Virginia, I came across a framed copy of an old map and a reference to Henry Sinclair's voyage of 1398 on the wall behind Leif's statue. In Halifax, at the Nova Scotia Public Archives, I discovered that Sinclair's saga had quietly surfaced some fifty years ago with the publication of a small booklet in the port town of Pictou. Its contents not only turned my attention away from New Ross to Guysborough Harbour and the province's north eastern shore, it also sent me exploring the life and times of the man described in the sixteenth-century Venetian document known as "The Zeno Narrative" as "a prince as worthy of immortal memory as any that ever lived."

4

A Northern Prince

Following a midday television interview that I gave about New Ross, I was contacted by Bill Sinclair, president of the Sinclair Clan in Canada. An amiable, retired career diplomat, he had called in following some comments made about the possible connection of the site to a Henry Sinclair voyage. A few days later I sat in his living room in Halifax, quietly sharing views not only about the feasibility of the voyage but also its possible connection to the Knights Templar and the legend of the Holy Grail. I also discovered that he and his brother Jack, a former senior federal bureaucrat with the ministry of health and welfare, were spearheading plans for the 1998 celebrations marking the 600th anniversary of the voyage.

The historical beginnings of the Sinclair family and its subsequent involvement with the Norwegian and Scottish courts stretch back to Viking times and the ninth century Norse invasion of the islands off the coast of northern Scotland where, as I found in my travels back to these ancestral lands, the name Sinclair seems rooted in the very rocks and as widespread as the wave-washed islands themselves. The family's original name was Moray, so called because one of their prominent leaders came from the region of Moeri in Norway. The Morays became the dominant Norse family of power and influence in these northern waters. An important early chapter in the family's history was also written in the more pastoral landscape of northern France, in the region that was to be known as Normandy, because of the same Norsemen who settled there.

Early in the tenth century a sword-slinging, free-booting Norseman known as Rolf the Ganger, a member of the Moray family, was exiled from Norway for an act of aggression that angered the king. In true Viking fashion, he took to sailing south in search of plunder and the possibility of a permanent place to settle. He found both on the coast of northern France. Having tried the patience and pockets of King Charles the Simple, with a persuasive Viking protection racket that would have done the Mafia proud, he was given a sizeable spread of real estate on either side of the Epte River. This historic deed was concluded at the Treaty of St. Clair in 911 whereby Rolf became the first Duke of Normandy. As was the case with Eric the Red, exile presented new opportunities. But unlike Eric, who hung on to belief in the Norse gods, Rolf converted to a new religion. Perhaps out of appreciation to the local saint after whom the castle of St. Clair-sur-Epte had been named, or the promise of marriage to one of the king's daughters, Rolf readily converted to Christianity. Consequently this marauding offspring of the Morays of Norway became transformed into a French St. Clair who, in the Latin vernacular of the ruling classes in Europe at the time, fathered the family of de Sancto Claro, the Holy Light. This name and the family's subsequent association with the Christian Crusades, the Knights Templar and the legend of the Holy Grail was to have a lasting influence and carry a deep symbolic significance for some of its later members in Scotland, where the name later changed to Sinclair.

A century and a half later some of Rolf's St. Clair descendants joined forces with the Duke of Normandy, known in England as William the Conqueror, as he set sail to begin the conquest of England following the death of Edward the Confessor. Opposed by the forces of his half-brother,

Harold, at the Battle of Hastings in 1066, he was ably assisted in his historic victory by a number of St. Clairs. Having courageously played its part in bringing Norman rule to England, the family was duly rewarded with the expected spoils of conquest in the form of possessions, power and prestige.

Meanwhile another Norman, William de Sancto Claro, had been placed in the service of the English-born Princess Margaret, whose brother Edgar had a more legitimate claim to the throne of England. Forced to take refuge in Hungary during this turbulent and dangerous time, Margaret and her brother were dutifully accompanied by her young St. Clair attendant. Then, when providence made Margaret wife to King Malcolm III of Scotland, young William accompanied her to her new-found domain in the north.

Apart from his official duty as cupbearer to the queen, he was also the protector of the Holy Rood, an exquisitely encased religious relic in the form of a piece of dark wood believed to have come from the cross on which Jesus died. Brought to Scotland from Hungary by the pious Margaret, it was considered to be one of the most sacred relics in all of Christendom, at a time when religious relics were highly esteemed and widely venerated. The deeply religious Scottish queen also had a small church built in honour of St. Katherine of Alexandria, in a valley surrounded by tree-covered hills, south of Edinburgh.

The lands through which the Esk River wound its way to the sea were also strategic to the defence of Edinburgh, and Scotland, from an English attack from the south. William was knighted by the Scottish king and he was given the lands of Rosslyn in perpetuity. By the end of the eleventh century the Sinclairs had become prominent players in national affairs, allied as they were with the Scottish monarchy. King Malcolm III had also forged a peaceful and purposeful alliance with his northern Norse neighbours, to whom his Sinclair vassal happened to be distantly related. It was an alliance that would thrust the Sinclairs into the position of international diplomats and would pave the way

Ruins of Rosslyn Castle, Scotland

Sketch of part of Rosslyn Castle courtyard.

for Henry's appointment, without any serious objection by the Scottish court, to the Norwegian earldom of Orkney.

During a second visit to Scotland in as many years, I ambled along the banks of the winding river where Henry Sinclair must have played as a boy and I sighted the remains of the once impressive Sinclair Castle where he was born. After climbing a winding path through forest and field, I stood in the castle's empty courtyard surrounded by crumbling walls. Here, as a young man, he had saddled up for the ride to Edinburgh in answer to a call of duty from the Scottish court. From here, he had set out to claim a Norwegian earldom in the far off Orkney Islands. Looking at the lingering remnants of this centuries-old castle, in a landscape that has changed little with the passage of time, it was easy to imagine the former magnificence of the place when it was the palatial home of one of the most prestigious families in all of Scotland.

Leaving by the narrow bridge over the moat, I followed a path to still higher ground to gaze at the intricate exterior of one of the most remarkable churches in all of Britain — Rosslyn Chapel, the Collegiate Church of St. Matthew, built in the mid-1400s by a member of the family of the Holy Light, the grandson of Prince Henry Sinclair. The highly

Rosslyn Chapel

ornamented chapel built by imported masons has been referred to as both a Bible in stone and the chapel of the Grail. It is also believed to contain concrete evidence of Henry Sinclair's voyage to North America. Having suffered harshly from the ravages of time, religious turmoils and the elements, as well as earlier attempts this century at preserving its uniquely carved interior, this chapel is now being properly protected from further decay and destruction and plans are in place to restore it to a semblance of its former glory.

By the beginning of the twelfth century the family had become significantly associated with the Christian cause at home and abroad. Apart from the symbolic spiritual significance of its Latin name, de Sancto Claro, still then in use in Scotland, the family performed the role of permanent protector of the relic of the cross of Christ. At a time of intense religious fervour, a time when members of noble families all across Europe and countless thousands of others responded to the call of Pope Urban II for a holy war against the Moslem controlled Middle East including Jerusalem and other sacred places of early Christendom, such a relic and those closely associated with it were held in the highest

North door - Rosslyn Chapel

esteem. The royally endowed symbol of the engrailed cross on their shield, the fluted cross of the St. Clairs, demonstrated to one and all their status as earthly guardians of a relic associated with the manifestation of divine love in the world.

Through the next two centuries, the Sinclairs remained closely associated with the Scottish crown. Consequently their holdings in Rosslyn and elsewhere increased with the good fortunes of such Scottish heroes as Robert Bruce. Unfortunately, Henry Sinclair's grandfather met his death while on another mission of sacred trust, to bury King Robert's heart in the Holy Land. In August 1330, along with the mission leader Sir James Douglas, he fell victim to a band of militant moors.

Born in 1345 to Isabella, daughter of the Norwegian-appointed Earl of Orkney, who had married Sir William Sinclair, Lord of Rosslyn, the young Henry enjoyed a privileged upbringing. It is very likely that he was told tales of the seafaring adventures of his Viking ancestors. He would have heard many stories, poems and songs that not only chronicled the accomplishments of his Norman and Scottish forefathers but which also heralded the heroic deeds of knights at arms and praised the practices of chivalry and courtly love. And as a future knight he would have been vigorously trained in sword play and the tactics of mounted combat.

The family's past involvement with the Crusades, beginning with the retaking of Jerusalem from Moslem control in 1099, and then with the Order of the Knights of the Temple of Solomon, whose Scottish headquarters at Balantrodoch were not far away from his birthplace, likely resulted in him being surrounded at a young age with talk of both the high idealism and the bloody reality of the cause for which some of his relatives had died. As he grew to manhood he must also have become acquainted with the military successes and failures and the almost unprecedented power and wealth of the Templar Knights. It is also very possible that he had heard of and perhaps even become personally

Ruins of Templar headquarters in Temple, Scotland, with Scottish Templar archivist Robert Brydon.

Templar grave slab at Temple, Scotland.

knowledgeable of the Templars' condemned secret rituals and unorthodox religious beliefs.

While driving the relatively short distance eastwards from Rosslyn to Balantrodoch, now called simply Temple, to walk among the ruins in the company of the loquacious Scottish Templar historian Robert Brydon, I could imagine the young Henry Sinclair riding across the same stretch of countryside.

With the death of his father, Sir William, in 1358, the thirteen-year-old Henry assumed the title Lord of Rosslyn. His future as a loyal servant of the Scottish crown seemed set out before him.

As we know, Henry Sinclair's lineage not only linked him to the Scottish court but also to that of Norway. Norwegian-appointed earls distantly related to the Sinclairs had ruled over the Orkney Islands, an archipelago of sparse and rugged islands lying north of Scotland, since the end of the ninth century when Norsemen had first settled there. But it was thanks to his mother that he had a legitimate claim to the Orkney earldom and was cast in the role of young diplomat by the Scottish king.

At the age of eighteen, he was knighted and appointed Scottish ambassador at the 1363 wedding of King Haakon VI of Norway to Princess Margrette, the ten-year-old daughter of the King of Denmark who would become the powerful queen of all of Scandinavia. In Copenhagen, where the wedding took place, Henry had the opportunity to press his claim to the Orkney earldom over two cousins who were already jockeying for the position. With his Orkney claim recognised by the Norwegian court, Sir Henry returned home to Rosslyn and later married Janet Haliburton, the reputed beauty of nearby Dirleton Castle. However, their early married bliss was short lived. With adventure in his veins and an obligation to answer a call to arms, he was soon saddling up for a military escapade far from home, one that was in keeping with the tradition established by several generations of Sinclairs.

Following his family's and his father's examples, Henry took up arms for a holy cause. In 1365 he joined a contingent of Scottish knights gathered in the port city of Venice in preparation for the crusade of Peter the First of Cyprus. The objective was to take and hold the Egyptian city of Alexandria, which was under Moslem control. Chartered Venetian vessels carried the crusading knights across the Mediterranean to their destination.

Historical documents of the period confirm that Carlo Zeno, the Venetian naval commander, who was to gain fame in Venice's dramatic victory over Genoa at the Battle of Chiogga fifteen years later, took part in this endeavour and that at the time he was introduced to a Scottish prince on his way to the Middle East. This suggestion of a meeting between Sinclair and Zeno likely led to the former using Venetian naval expertise in his successful effort to exert complete control over his scattered island domain.

It is known that although the crusade that Henry Sinclair was a part of failed to hold Alexandria against a counter attack, he was able, thanks to earlier negotiations, to gain a guarantee of safe passage as a pilgrim to Jerusalem. For such a purpose the word of the Moslem overlords could almost always be trusted for, as devout worshippers of the One God, they respected the path of pilgrimage. Henry visited and prayed in the most sacred shrine in all of Christendom, the Church of the Holy Sepulchre, believed to have been built over the spot where Jesus had been both crucified and buried, again thanks to the tolerance of the Islamic rulers. This pilgrimage earned him the nickname of Henry the Holy among a few cynics at home. However, his involvement in the crusade of King Peter of Cyprus gained him respect where it mattered. Far from becoming a reclusive, pious knight on his return to Scotland, he was given command of a Scottish force that invaded northern England. His proven service to cross and crown, his experience on the field of battle at home and abroad, his diplomatic skills and his family's long history of service to the Scottish monarchy led to his appointment by King David II as Lord Chief Justice of Scotland and Admiral of the Seas.

In 1379, King Haakon VI of Norway, aware of Henry's Norse lineage and his rightful claim and perhaps also impressed by his military prowess and his family's wealth, finally agreed to invest him as the legitimate Earl of Orkney over the claims of his rival cousins. The Lord of Rosslyn certainly had all the makings of a man capable of controlling Orkney and carrying out the wishes of the Norwegian king. However, since Henry was also a dedicated vassal of the Scottish monarch and since the Orkney earls had a reputation for being a somewhat independent breed, Haakon took the precaution of attaching some very stringent conditions to the title. Apart from the customary up-front contribution in hard cash, Henry was required to swear to defend the King of Norway's interests in the islands and neighbouring regions at all times and to aid the king in his wars by supplying a hundred men at arms on due notice. He was to continually protect the interests of the islanders and not sell or pledge off any part of the island kingdom. And by way of keeping him in line, he was forbidden to build a fortification there without the king's express permission.

In a move seen by some as an indication that he may have put his own political interests ahead of those of the Norwegian monarch, he proceeded to build himself a strong coastal fortress in Kirkwall. It was so strong that it took many a cannon ball and explosive charge to breach its walls over two and a half centuries later. Sinclair's harbour fortress stood out among the other buildings in Kirkwall as testimony to his determination to maintain his rule against any who would or could challenge him. Behind its massive walls he established his court where, according to Sinclair family historian, Fr. Richard Augustine Hay, he at times maintained "three hundred men with red scarlet gowns and coats of black velvet." Other than Castle Street in central Kirkwall and an heraldic stone embedded in the wall of a local building, nothing remains today of this formidable fourteenth-century edifice. Dr. Peter Anderson, Scottish his-

Ruins, pre-dating the Viking period, at Skara Brae, Orkney.

torian and deputy keeper of the Scottish Record Office, whom I met during a conference in Kirkwall, provided me with a detailed outline of the waterfront castle's configuration and a drawing of what it is believed to have looked like.

Henry's investure, in spite of its strictures, made him a virtual prince in this island kingdom. He had been given the power to make laws and wore a crown when he did so. He had the ceremonial sword of state carried before him. He could issue coin of the realm. He could remit crimes, and with adequate manpower and a sizeable fleet, he was able to subdue any threat to his control of the islands and their resources, especially that of the powerful, money-siphoning, English-appointed Bishop of Orkney. Ready to take on the powerful bishop, the remnants of whose palace still stands in the centre of Kirkwall on the other side of the impressive eleventh-century St Magnus Cathedral, Henry learned that the hated cleric had become the sorry victim of a people's

Approaching Orkney

uprising. The islanders were in all probability encouraged by the presence of the Sinclair earl. In true feudal tradition, in 1384, a Sinclair was appointed to the Orkney bishopric.

According to Hay, Henry Sinclair came to be more honoured than any of his ancestors. Hay also states that he was considered to be second only to the King of Norway, Sweden and Denmark, whose court he attended on a number of occasions as elector and vassal. He had risen to an enviable position of power. According to Dr. Anderson, Henry Sinclair, although never officially designated the title of prince, was unquestionably the King of Norway's single most powerful subject.

We know that in 1389 he was in Norway at the investure of young Eric of Pomerania as the chosen successor to the Scandinavian throne. But the real power lay in the hands of the regent, the capable Queen Margrette whose wedding Henry had attended many years earlier in Copenhagen. As

The arms of the Earl of Orkney, almost certainly those of Henry Sinclair.

he had been to her husband, so he remained to her, a trusted and powerful ally in the struggle to preserve Nordic control in northern seas, a matter that may also have influenced his decision to sail westwards in 1398.

His coat of arms now displayed a Viking style sea dragon atop the crowned head of an armoured knight, and underneath, a shield emblazoned with the engrailed cross of the Sinclairs. It reflected his status as a Norse earl of an island kingdom, the regal status of his position as well as his Scottish family's hereditary role as keepers of a sacred trust. To prove his mettle he used his considerable resources to fit out and man a fleet of ships capable of maintaining supremacy in those often turbulent and treacherous seas.

Having experienced a howling, gut-wrenching storm in the same waters on my return from Orkney to the north coast of Scotland, a storm that tested the ability and perseverance of a modern ship's equipment and crew, I gained a healthy respect for the task Sinclair and his men undertook in their much smaller and more exposed wooden vessels.

Ready to expand his island empire, presumably with the blessing of the Norwegian crown, and having already claimed jurisdiction over the Shetland Islands to the north after killing his rival Malise Sperra in 1390, Sinclair now undertook to bring the more westerly Faeroe Islands, with their resources of wool and fish, under his control. It was this thrust out into the western Atlantic that brought Sinclair either by design or destiny, in contact once again with a member of the Zeno naval family from Venice.

5

A Manuscript and Map

Around the year 1520, Nicolo Zeno, a young boy playing in his family palatial home in Cannaregio in Venice, carelessly tore up a collection of old letters and charts he had come across in an unlocked chest. As he later learned, most had been written and drawn by his illustrious ancestors, Antonio and Nicolo Zeno, sometime between 1390 and 1404.

However, his developing an interest in history prompted him to search out the remnants of the letters and charts. After frantically reassembling their torn and faded remains, he discovered that the letters told of voyages across the North Atlantic embarked on by his relative Antonio while he was in the service of a prince who ruled an island kingdom north of Scotland. They had been sent back to an older brother Carlo, in Venice. Although difficult to read and decipher, Nicolo, the historian, was able to copy enough material from the fragments of the letters to assemble and publish in 1558 the historic document known as the Zeno Narrative.

Nevertheless, the first published reference to this voyage occurred a little earlier, in a publication called *Discendenze Patrizie*, a kind of encyclopaedia of notable Venetians written by historian Marco Barbaro in 1536. The original, which

DE I COMMENTARII DEL
Viaggio in Persia di M. Caterino Zeno il K.
& delle guerre fatte nell' Imperio Persiano,
dal tempo di Vssuncassano in quà.
LIBRI DVE.
ET DELLO SCOPRIMENTO
dell' Isole Frislanda, Eslanda, Engrouelanda, Estotilanda, & Icaria, fatto sotto il Polo Artico, da due fratelli Zeni, M. Nicolò il K. e M. Antonio.
LIBRO VNO.
CON VN DISEGNO PARTICOLARE DI
tutte le dette parte di tramontana da lor scoperte.
CON GRATIA, ET PRIVILEGIO.

IN VENETIA
Per Francesco Marcolini. M D LVIII.

Title page of the original Venetian edition of the Zeno Narrative, 1558

is presently in the Correr Museum in Venice's Piazza di San Marco, contains a reference to the Zeno family, a family noted for its service at sea and in diplomacy and includes the following reference, which in its English translation reads: "Nicolo the Chevalier [was] in 1379 captain of a galley against the Genoese. [He] wrote with his brother [of] the voyage to Frislanda, where he died. Antonio [also] wrote with his brother Nicolo the voyage to the islands near the Arctic Pole, and of their discoveries of Estotiland in North America. He [Antonio] remained fourteen years in Frislanda; that is four with his brother and ten alone."

The publication 20 years later of the rewritten contents of the Zeno Narrative was a detailed and dramatic account of Antonio's voyage to North America and also the interesting events leading up to it. The accompanying map indicated knowledge of lands lying to the southwest of Greenland, referred to as Estotiland and Drogio. These have been identified from time to time as Newfoundland and Nova Scotia.

In spite of its historical contents, apparently few people in Europe in the mid-sixteenth century showed much interest in the contents of the Zeno publication. It suffered from being presented as a family history with the rather pedantic and obscure additional title, *The Discovery of the Islands of Frislanda, Eslanda, Engronelanda, Estotilanda and Icaria; Made by Two Brothers of the Zeno Family*. This was a Europe still buzzing with the accounts and experiencing the dramatic political and commercial consequences of Columbus's well-publicised transatlantic voyages. However the Zeno map, with its accurate depiction of the coastline of Greenland, was used to good effect by such notable northern mariners of the time as Martin Frobisher and John Davis. In spite of the geographic and historic details contained in the Zeno publication, the identity of the "great lord" who ruled islands north of Scotland remained a mystery, that is, until a German-born navigational historian, John Reinhold Forster, took an interest in the Venetian manuscript and in 1784 published his *History of the Voyages and Discoveries Made in the North*. Forster not only identified the strangely named northern islands as being the Faeroes, Iceland and Greenland, but he also got a fix on the man whose service Antonio was in. Historical records suggested to Forster that he must have been the Scottish-born Henry Sinclair, Lord of Rosslyn, who as the Earl of Orkney during the latter half of the fourteenth century also controlled the Faeroes and had contact with Iceland and Greenland.

It was almost another hundred years before the Zeno narrative was translated into English. Richard Henry Major, Secretary of the Royal Geographic Society and Superintendent of the map room at the British Museum, was commissioned in 1873 by the renowned Hakluyt Society to translate the work. The society, which stemmed from the diligent record keeping of the geographer and marine historian Richard Hakluyt in the Elizabethan period, had been formed in 1846 to preserve and promote knowledge of early transatlantic adventures and expeditions. The translation made fascinating reading and almost immediately gained the Sinclair voyage many supporters in the English speaking world.

After praising the wide ranging accomplishments of relatives from the year 1200, Nicolo Zeno gave a brief introductory account of the events leading up to his ancestors' adventures:

> Now M. Nicolo the Chevalier, being a man of great courage.....conceived a very great desire to see the world and to travel and make himself acquainted with the different customs and languages of mankind, so that when occasion offered, he might be the better able to do service to his country and gain for himself reputation and honour. Wherefore having made and equipped a vessel from his own resources, of which he possessed an abundance, he set forth out of our seas, and passing the Strait of Gibraltar, sailed some days on the ocean, steering always to the north, with the object of seeing England and Flanders.

It is a historical fact agreed to even by those who doubt the veracity of the entire document that voyages from Venice to Flanders and other northern ports took place at least annually from the early 1300s onward. Nicolo's voyage, which possibly took place in the 1380s, was not unusual. Of course, in view of the fact that he was a member of a Venetian family involved in commerce, he may well have set out with the intention of making secret contact with a potential trading partner in lucrative northern waters, waters increasingly frequented by the ships of the German Hanseatic League, a consortium of aggressive German merchant cities.

The Major translation of the Zeno document explains that the ship was caught up in a fierce storm in the English Channel and was tossed about for many days. Nicolo and the men with him had no idea where they were even when they sighted land, which in this instance turned out to be an island called Frislanda. Based on Forster's geographical research it was clearly identified in the translation as the main island of the Faeroes. Although the ship was wrecked on the rocky shore, the Venetian crew managed to survive. But their troubles were not over yet. As was often the practise in these rugged and desolate North Atlantic islands, some of the locals attacked the sailors as they were struggling to reach the safety of dry land, with the intent of robbing them of their few possessions and claiming any cargo and wreckage they could find. The much weakened Venetians tried desperately to defend themselves in the fracas. Fortunately their cries and yells brought timely intervention from a chieftain in command of an armed retinue who happened to be on the island. On hearing and seeing the cause of the commotion, he set his men upon the attacking islanders. After the remaining locals had been subdued, this chieftain addressed the shipwrecked crew in Latin asking them who they were and where they came from. On learning that they were Venetians, the Zeno account tells,

> [after] promising them all that they should receive no discourtesy, and assuring them that they were come into a place where they should be well used and very welcome, he took them under his protection, and pledged his honour for their safety. He was a great lord, and possessed certain islands called Portlanda, lying not far from Frislanda to the south, being the richest and most populous of all those parts. His name was Zichmni, and besides the said small islands, he was Duke of Sorano, lying over against Scotland.

Of course, like Forster, Major also concluded that this great lord, whom Nicolo Zeno called Zichmni, was Henry Sinclair, the powerful earl of the Orkney Islands, who was in the process of subduing the Faeroes. The reference to him being "Duke of Sorano, lying over against Scotland" also fits into the Sinclair profile since the Orkney earldom also included the lands known as Caithness, in northern Scotland.

On learning that Nicolo was the brother of Carlo, a renowned Venetian naval commander, Sinclair had him and his men join his own fleet of thirteen vessels, with Nicolo serving as advisor to the commander. There is documented evidence to suggest that Sinclair had met Carlo in Venice while on his way to the Middle East as a crusader. There can be no doubt that Sinclair must have been impressed by the ships, the naval fire power and expert seamanship of the Venetians during his voyage across the Mediterranean. Finding a group of Venetian sailors on his doorstep, he was quick to befriend them and, as the Zeno Narrative glowingly recounts, the superior navigational skills of the Venetians did result in military triumphs for the Orkney earl throughout these rugged islands. They also brought Nicolo and his men honours and rewards:

> The Venetians received from all such great honour and praise that there was no talk but of them, and of the

> great valour of Messire Nicolo. Whereupon the chieftain, who was a great lover of valiant men, and especially of those that were skilled in nautical matters, caused Messire Nicolo to be brought before him, and having honoured him with many words of commendation, and complimented his great zeal and skill...he conferred on him the honour of knighthood and rewarded his men with many handsome presents.

At this juncture we learn that Nicolo, now fully in command of the Sinclair fleet, suggested in a letter to Antonio, his younger brother back in Venice, that he should also outfit a vessel and join him in this enterprise in these northern islands where, he pointed out, fish were so plentiful that a fortune was to be had shipping them to markets in Flanders, Brittany, England, Scotland, Norway, and Denmark. In about 1390, Antonio joined Nicolo in the Faeroe Islands, not expecting that it would be at least fourteen years before he would see his beloved Venice again.

The year after Antonio's arrival, no doubt under Sinclair's orders, Nicolo set sail, "with a view of discovering land," in the direction of Iceland and Greenland, where the Norse populations had been reduced by that time. Famine, disease, the unforgiving landscape, natural disasters and adverse climatic conditions had taken their toll in both countries.

Henry Sinclair's Norwegian lineage certainly put him in the position of knowing about the Viking voyages of discovery of a few centuries earlier. The ongoing but infrequent trading runs between these northern domains must certainly have kept this lore alive. In view of the devastating effects of the plague, which spread northwards from Europe, a voyage to explore the living conditions in those sparsely inhabited lands or to find the even more distant lands to the west would have made sense at this time. So whether it was for the purpose stated, or for commercial reasons, or as part of a politically motivated voyage, Nicolo fitted out three small ships and headed out into the dangerous waters of the North Atlantic. He seems to have reached the northwest coast of Greenland without difficulty. There, as the Zeno Narrative so descriptively informs us, he found an extraordinary monastery close to a hill "which vomited fire like Vesuvius or Etna." The monks "of the Order of Friars Preachers" had industriously harnessed the hot springs close by and had devised ingenious ways of using the hot water for heating and cooking and also in their enclosed gardens, which produced flowers, fruit, vegetables and herbs. The hot volcanic rock dissolved into a very workable white lime when cooled quickly with water. With this material the monks were able to create buildings of appealing design. Both geological and archaeological research carried out in the twentieth century on the northeast coast of Greenland have shown that active volcanoes and hot springs once existed near the remains of St. Olaf's monastery, in the vicinity of Gael Hamke Bay.

After describing the good trading and working relationship between the native population and the friars, the Zeno Narrative gives an accurate description of the buildings in which the local Inuit lived and the boats they used for fishing.

> Their houses are built about the hill on every side, round in form, and twenty-five feet broad, and narrower and narrower towards the top, having at the summit a little hole, through which the air and the light comes into the house; and the ground below is so warm, that those within feel no cold at all ... The fisherman's boats are made like a weaver's shuttle. They take the skin of fish and fashion them with the bones of the self same fish and, sewing them together and doubling them over, they make them so sound and substantial that it is wonderful to see how, in bad weather they will shut themselves close inside and expose themselves to the sea and the wind without the slightest fear of coming to mischief. If they happen to

> be driven on any rocks they can stand a good many bumps without receiving any injury. In the bottom of the boat they have a kind of sleeve, which is tied fast in the middle, and when any water comes into the boat, they put it into one half of the sleeve, then closing it above with two pieces of wood and opening the band underneath, they drive the water out; and this they do as often as they have occasion, without any trouble or danger whatever.

Unfortunately Nicolo died soon afterwards of an illness brought on by the severe cold and hardships experienced on this northern voyage. Antonio, who quite understandably would rather have returned to the comfort and warmth of his native Venice, was given all of Nicolo's accumulated wealth and honours and encouraged or enticed by Sinclair to stay in his northern island kingdom. It was Antonio's subsequent series of letters back to Carlo in Venice that provided the details of what later became identified as the Sinclair transatlantic voyage of 1398 and the beginnings of the Sinclair saga.

As with many of the Viking sagas, the Sinclair saga typically began with a story of a ship being blown off course by a storm, onto the shores of an unknown land to the west. At first the plan was for Antonio to set out with a few vessels in search of a land which a fisherman had excitedly talked about as "a New World." Zeno gave this colourful account of the background to the historic voyage.

> Six and twenty years ago four fishing boats put out to sea and encountering a heavy storm were driven over the sea in utter helplessness for many days; when at length, the tempest abating, they discovered an island called Estotiland, lying to the westwards about one thousand miles from Frislanda. One of the boats was wrecked and six men that were in it were taken by the inhabitants, and brought into a fair and populous city,

> where the king of the place sent for many interpreters, but there were none could be found that understood the language of the fishermen, except one that spoke Latin, and who had also been cast by chance upon the same island. On behalf of the king he asked them who they were and where they came from; and when he reported their answer, the king desired that they should remain in the country. Accordingly as they could do no otherwise, they obeyed his commandment and remained five years on the island and learned the language. One of them in particular visited different parts of the island and reports that it is a very rich country, abounding in all good things. It is a little smaller than Iceland, but more fertile; in the middle of it is a very high mountain, in which rise four rivers which water the whole country.
>
> The inhabitants are very intelligent people and possess all the arts like ourselves; and it is believed that in time past they have had intercourse with our people for he said that he saw Latin books in the king's library, which they at the present time do not understand. They have their own language and letters. They have all kinds of metals, but especially they abound with gold. Their foreign intercourse is with Greenland, whence they import furs, brimstone and pitch ... They sow corn and make beer, which is a kind of drink that northern people take as we do wine. They have woods of immense extent.

Although it was described as being about 1600 kilometres west of the Faeroes, this was very likely an underestimated assessment of the distance travelled due to the storm. The Zeno map showed a portion of the coastline of Estotilanda and it bears a remarkable similarity to part of the shoreline of northern Newfoundland.

The fisherman's account continues with his adventures in more temperate lands to the south. He and his companions had set out for a land named Drogio. After riding out a storm they arrived in an unknown country where the natives killed and ate many of them. Fortunately a few were spared because they managed to show these naked cannibals how to fish more effectively with nets. Word of their fishing skills spread to other hostile tribes. After spending time with several of these people, they made their way across country, to the southwest, where they discovered a civilisation with cities and temples, but which also practised human sacrifice.

From these descriptions it would seem that after leaving Estotilanda, their sea journey took them all the way south to the coast of Central or South America and that they then travelled inland. After many years in the southern hemisphere, this fisherman managed to make his way back by land to Drogio. From there he succeeded in getting passage on a trading ship to Estotilanda. He put his new-found knowledge of tribal languages and the geography of the continent to the south to good effect with passing traders so that after three more years he was wealthy enough to fit out a seagoing vessel. With a few other cast-off sailors who had also been living in Estotilanda, he sailed back across the North Atlantic to his original home, where he told his remarkable tale.

To an island people who had already heard many stories of unusual lands beyond the western horizon, his story was quite plausible. When it reached Sinclair's ears, he immediately decided to send Antonio and a few ships westward in search of these lands. The intended expedition aroused such widespread interest that there was no problem getting the manpower or resources together. Even though the fisherman, who was to be their chief guide, died before departure, others who had returned back across the Atlantic with him became part of the expedition, which we are told included a considerable number of ships and men. When it finally set sail from either the Faeroe or the Shetland islands, it was commanded not by Antonio but by Sinclair himself whose sturdy ship showed the eight-pointed Templar cross on its foresail.

Not surprisingly the expedition was hit by a howling storm which blew for eight long days and resulted in the loss of a considerable number of boats in the fleet. When the storm abated the scattered survivors had no idea where exactly they were on the vast expanse of ocean. After finding each other again they continued sailing westwards, helped by a favourable wind, and eventually sighted land.

Although desperate to replenish their supplies of wood and water, they hesitated to go ashore because of the threatening signals of a large party of armed men who showed up on the headland. Their actions and yelling were perceived as unwelcoming gestures. A former Shetlander who had lived among the natives for several years explained that they would only allow one man ashore to live among them. After finding a small natural harbour on the east coast of what seemed to be a large island, a party of sailors landed with the intention of quickly gathering supplies. A group of armed natives suddenly appeared out of the woods and made their earlier threats all too tragically real. Several of the men were killed.

The saga goes on to say that frustrated in his attempt to secure a safe landing place,

> [Sinclair] took his departure with a fair wind and sailed six days to the westwards; but the wind afterwards shifting to the south west, and the sea becoming rough, we sailed four days with the wind aft, and at length discovering land, as the sea ran high and we did not know what country it was, were afraid at first to approach it; but by God's blessing the wind lulled, and then there came on a great calm. Some of the crew then pulled ashore, and soon returned to our great joy with news that they had found an excellent country and a still better harbour. Upon this we brought our barks and our boats to land, and on entering an excellent harbour, we saw in the distance a great mountain that

> poured forth smoke, which gave us some good hope that we would find some inhabitants in the island.

Sinclair immediately sent a reconnaissance force of 100 soldiers inland in the direction of the smoking mountain, to check out the native inhabitants. The rest of the expedition quickly gathered wood and water, and were delighted to be able to catch "a considerable quantity of fish and sea fowl" and "such an abundance of birds' eggs, that our men, who were half famished, ate of them to repletion."

Soon after the ships dropped their anchors, the month of June arrived and the men, more accustomed to the brisk and briny atmosphere of North Atlantic islands, found the air mild and pleasant beyond description. The same observation had been made centuries earlier by the Vikings who settled in Vinland and was again, centuries later, by the first French and English settlers on arriving in Nova Scotia.

According to Zeno, Sinclair christened the natural harbour where they had anchored Trin Harbour and then gave the name Cape de Trin to the long headland leading out to the sea.

Eight days after setting out on their trek into the interior, the foot-weary men returned with the news that they had discovered the cause of the smoke. It came from a great natural fire which burned at the bottom of a hill where there was a spring from which flowed a steady stream of a pitch like substance. This black stream ran all the way to the sea. Nearby they saw a large river and another very good and safe harbour. They also reported that they had seen a large group of natives who were small in stature, seemed timid and hid in caves on seeing the newcomers. The saga goes on to say that when Sinclair heard this "and noticed that the place had a wholesome and pure atmosphere, a fertile soil, good rivers, and so many other conveniences, he conceived the idea of fixing his abode there and founding a city."

However, many of his men, having been through a rough voyage and concerned that if they stayed too much longer

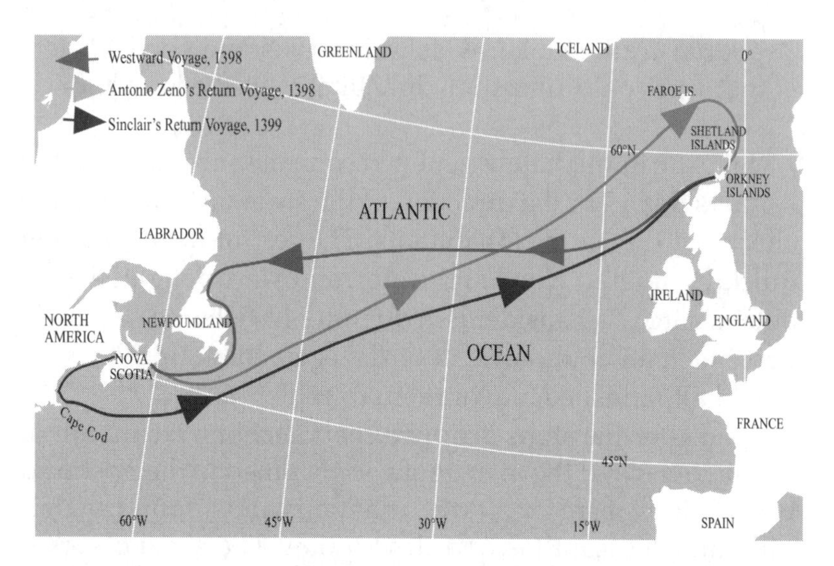

Prince Henry Sinclair's transatlantic expedition 1398-1400 from the Orkney Islands to the New World, and Antonio Zeno's return in 1398.

they might not be able to sail back home until the following summer, murmured their disapproval. From part of Antonio's final letter sent back to Venice following his return to the Faeroe Islands, we learn that his commander "retained only the row boats and such of the people as were willing to stay with him, and sent all the rest away in ships, appointing me, against my will, to be their captain. Having no choice therefore I departed and sailed twenty days to the eastwards without sight of any land; then turning my course towards the south east in five days I sighted on land, and found myself on the island of Neome, and knowing the country I perceived that I was past Iceland; I took in fresh stores and sailed with a fair wind in three days to Frislanda. Here the people, who thought they had lost their prince, in consequence of his long absence on the voyage we had made, received us with a hearty welcome."

The Sinclair saga closes with the comment from its sixteenth-century compiler, culled from the remains of Antonio's last letter, that the northern prince after first settling down in the "excellent harbour" of his newly discovered land then set out to explore the whole of it.

While doing some research at St. Mary's University library, in Halifax, I found a copy of *The Sinclair Expedition to Nova Scotia in 1398*, published by the *Pictou Advocate* in November 1950. Written by Fredrick J. Pohl, it claimed that, based on extensive field research, the Sinclair expedition's landfall was in northeastern Nova Scotia. A year later I drove up along the province's eastern shore to see for myself this "excellent harbour" and to walk the terrain where Sinclair and his men were believed to have first come ashore.

6

An Excellent Harbour

"Perhaps you'll get to see Wilma the whale," said Don Colp, an active member of the Guysborough Historical Society as we walked out to the mouth of Guysborough Harbour. His comment caught me momentarily by surprise, preoccupied as I was with the possibility that I was walking over the same ground where Sinclair had first come ashore, the ground where perhaps a few fourteenth-century perpetuators of the outlawed European order of the Knights Templar had brought themselves to a new continent. Hearing Don mention a whale again I suddenly recalled the media stories about the young female Beluga that had shown up at the mouth of Guysborough Harbour a year or two earlier. Playfully friendly, Wilma had decided to stay within sight of the shoreline where Don and his wife Janelle were her nearest human neighbours. Although I did not get to see Wilma, I did get a close-up view of the layout of the natural harbour with its narrow mouth and protective sand bar, details which contributed to the conclusion arrived at by Frederick J. Pohl, the author of several books on early trans-atlantic voyages, that Guysborough had been the "excellent harbour" mentioned in the Zeno Narrative.

My introduction to Pohl's work had come at a conference I had organised in Nova Scotia about Oak Island. A professor of English literature with a passion for Shakespeare, Pohl was also a prolific writer. Apart from being a playwright, he had also devoted a good deal of his time and talents to writing about transatlantic exploration. His critically acclaimed, insightful biography of Columbus presented the man and mariner in a whole new light. Highly regarded also was his book about the sixteenth-century explorer Americo Vespucci, whom some people believe gave his name to the continent, although he named it Arcadia after the pastoral paradise of Greek legend. But long before any official confirmation of the fact, Pohl also pursued and popularised a hypothesis that others had arrived in North America long before Columbus. Pohl was author of such books on the topic as *The Vikings on Cape Cod, Atlantic Crossings Before Columbus* and *the Viking Settlements in America.* Like his contemporary in this field, Dr. Barry Fell of Harvard, Pohl was a maverick and a relentless historical sleuth who, whenever possible, literally walked the ground he was writing about. Neither did he hesitate to sail beyond established academic thinking in his search for answers. Apart from having been the first North American writer to give voice to the notion that Henry Sinclair had crossed the Atlantic almost 100 years before Columbus, he had also succeeded in adding some solid substance to the theory. Like most historians, he drew from the collected work of others, as I was to discover while reading through volumes of his letters and papers in the Public Archives of Nova Scotia.

I had been made aware of this interesting collection by Donald Bird of Truro, who had been the provincial planner for Nova Scotia. He had first become aware of Pohl's work while carrying out his own private research following the discovery of a large man-made stone structure near Halifax airport some years ago. Made up of a mammoth granite boulder balanced on smaller ones it was somewhat similar in design to ancient stone structures found in Ireland, Scotland and elsewhere. After corresponding with Barry Fell, he came to believe in the possibility of pre-Columbian visitors to North America. Then, after reading of Pohl's 1950 field research in the Guysborough and Pictou regions of Nova Scotia, he also began to take seriously the theory that a fourteenth-century Scotsman had successfully led a transatlantic expedition to these shores. On learning of my intention to write a book on the subject, Bird, a member of the New England Antiquities Research Association, trustingly mailed me large manila envelopes full of relevant articles, notes and correspondence.

Pohl was born in Westfield, Massachusetts and died in 1991 at the well-seasoned age of 101. He taught English at Ohio State University and at several other colleges and schools. Over the years his interest in early transatlantic exploration grew and resulted in the volumes already mentioned. His work earned him many honours, among them recognition from the Institute for the Study of American Cultures.

His writings were always the result of a refreshing open-mindedness coupled with exhaustive and methodical research, an approach I had come to appreciate both for the detailed work involved and the sometimes surprising and rewarding results it could bring. His lifelong search for traces of historic truth buried in the mysteries of the past left a rich legacy that has increased incrementally with such discoveries as the Viking settlement at L'Anse aux Meadows, Viking artifacts in New England and elsewhere, and, of course, the more recent revelations and the ever-increasing interest surrounding the Sinclair voyage.

Always fascinated by tantalising references in old documents and maps suggesting that North America had been visited by Europeans prior to Columbus, Pohl was naturally intrigued by the contents of the Zeno Narrative and the case that might be made in favour of the Sinclair voyage, if only some supportive evidence could be found. And, as sometimes happens in scientific or historical research, that is exactly what another enthusiast had just done.

The reference in the Venetian document to the discovery by Sinclair's men of a fire at the base of a high hill from which an open stream of pitch-like substance flowed into the sea had prompted Dr. William Hobbs, an American geologist, to investigate where such natural phenomenon might have existed at the time. He concluded that the place in question had to have been in the hilly, coal-rich Stellarton area of Nova Scotia, just south of Pictou, where burning open coal seams would have resulted in these conditions. He backed up his conclusion by pointing out the fact that no such juxtaposition of geological and geographical phenomena existed anywhere else within an eight-day, two-way foot march of the Atlantic coast of North America. Therefore, Hobbs deducted, Sinclair's landfall had to have been somewhere along the shore of northeast Nova Scotia. He suggested Tor Bay which lies southwest of Cape Canso. It was an impressive piece of scientific detective work and after hearing Hobbs expound on this and other facts related to the Sinclair voyage at the Explorers Club in New York, Pohl decided to spend some of the summer of 1950 in Nova Scotia exploring the terrain for himself.

During his initial research visit to the province, sometimes wearing a Sinclair Clan tie for luck, Pohl met with Arthur Godfrey, the editor of the *Pictou Advocate*. According to the Pohl letters in the Nova Scotia archives both men shared a number of interests, including philosophy, politics and, of course, history. At Pohl's request and expense, Godfrey agreed to publish a booklet about the Sinclair voyage which would also contain the results of his Nova Scotia research, especially his hypothesis that Sinclair had not sailed into Tor Bay but into the wide expanse of Chedabucto Bay, further up the coast. Cape Canso matched the Zeno reference to "the headland which stretched out into the sea." Pohl then reasoned that the "excellent harbour" in which Sinclair had dropped anchor must have been at Guysborough where the narrow entrance to the north and the long, protective gravel bar created a good natural harbour.

Pohl then suggested that the new arrivals would have climbed the highest hill in the area, a normal practice for explorers in new terrain, which happened to be Salmon Hill,

Entrance to Guysborough Harbour showing protective bar.

not too far to the south of Guysborough, from which they would have had a commanding view of the interior of the countryside. From this vantage point they would have been able to see smoke rising from behind hills lying to the west. At least one of these hills, the highest, was directly in line with the Stellarton region where, as the geologist Hobbs had pointed out, Sinclair's men would have discovered the smoke was caused by burning coal seams. The discovery by Sinclair's men of "a large river and a very good and safe harbour" in the same region was, according to Pohl, a reference to the nearby East River and to Pictou Harbour into which it flowed.

Pohl also succeeded in discovering the most likely year in which the expedition arrived in Nova Scotia. It had to have been sometime between 1396 when Sinclair was likely at Rosslyn and 1403 by which time he was believed to be back in Orkney. After wondering why Sinclair had given the name Trin to the "excellent harbour," it dawned on Pohl that explorers of the time often gave the name of saints or church holidays to their discoveries, a practise maintained through ensuing centuries. It struck him that the name Trin may have been an abbreviation by Zeno for Trinity. Zeno also mentioned that soon after the expedition dropped anchor "the month of June came in" and Pohl found from Vatican records that the year in which Trinity Sunday fell closest to the beginning of June was 1398.

Pohl's final contention in the 1950 booklet was that the rowboats Sinclair retained were likely of Norse design and quite large. Having a stepped mast and up to 30 oars they would have been ideal for coastal exploration. He also suggested that Sinclair, expecting supply ships to arrive the following year, most likely established his first settlement on the strategic high ground just inside the entrance of Guysborough Harbour, the same site later occupied by the seventeenth-century French explorer, trader and writer, Nicolas Denys, and on which a fort was also built.

Soon after reading about Pohl's conclusions, I found myself scanning a map of Nova Scotia anxious to take a trip to see the locations for myself. In the summer of 1996 I set out on a province-wide reading tour connected with my book on Oak Island. The tour turned out to be a most enjoyable and worthwhile experience. As luck would have it, my final stop was the library in New Glasgow which is quite close to both Stellarton and Pictou.

On the ride up from Mahone Bay, I was pleased to have the company of my son Aengus who was visiting from Ontario. A singer and song-writer, he was on his way to Prince Edward Island to meet up with a friend and to do some busking. We had not seen each other since he headed north a year earlier so it was a welcome time together, travelling through the Nova Scotia countryside.

We spent our last evening in the pleasant, waterside town of Pictou, which had once been a major encampment site for the Mi'kmaq who came to the shores every spring and summer to feast on the plentiful supply of clams and fish available in the river mouth and harbour area. In fact the name Pictou originated with the Mi'kmaq and has been said to mean either a place of fire or an enclosed bay. Pictou was also where the Mi'kmaq first met with their god-like hero Glooscap who, according to Mi'kmaq oral history, came across the sea from the east. The first people of European origin accredited with being in the Pictou area were believed to have been French fur traders who made their way there in the early 1600s. There is some archaeological evidence that French families built homes near Pictou and New Glasgow later in the same century. Of course, given its geographic location on the Northumberland Strait, which may have been sailed into by the French explorer Jacques Cartier, it is also feasible that Breton and Basque fishermen came ashore in this secluded harbour even earlier.

The first officially recognised settlement in Pictou came about following the ascendancy of the British in the colony, after the final defeat of the French at Louisbourg in Cape Breton. A large land grant involving several hundred thousand acres was issued in 1765 to what was known as the Philadelphia Company. The company, which included the renowned American inventor, newspaper publisher and politician Benjamin Franklin, sent about 35 settlers, six families in all, and supplies, up the coast by ship from Maryland.

The more popularly promoted history of the settlement of Pictou highlights the arrival of some 200 Scottish Highlanders who arrived six years after the Marylanders on board the *Hector*. After a gruelling 11-week crossing on the aging, Dutch-built ship, during which time hunger and disease took their toll and a fierce storm off the south coast of Newfoundland almost put a watery end to their hopes and dreams, the weary Scottish settlers arrived in Pictou Harbour on September 15, 1773. In reality their troubles were only beginning for they faced a situation for which they were totally unprepared. Finding that the preferable waterfront lands were already taken, in part by the accompanying agent John Ross, a fellow Scot, they were offered land deep in the forest away from the rivers and shoreline. There was also a serious lack of provisions necessary to sustain them during the winter so that they were compelled to borrow and barter and even had to raid the company store in an attempt to feed their families. Disillusioned and desperate, some of these unfortunates took off for Truro and Halifax where they hoped to earn their livings. A few returned the following year determined to make the best of a bad situation.

Such were the inauspicious beginnings of what was to become a veritable influx of Scottish settlers into that part of the province during the ensuing years. The cultural seeds planted by the *Hector* highlanders have flourished into a lively tradition of Scottish music and other events, such as the Highland Games. Aengus and I enjoyed a performance by local musicians in Pictou that attests to this enduring link.

The next morning, after leaving Aengus at the nearby ferry terminal, I gave a reading at the New Glasgow library and answered questions from the audience of eager young listeners. These readings invariably attracted two or three interested adults as well and so I soon found myself discussing not only the Oak Island mystery but also the Sinclair voyage. From one of them I learned that there were already tentative plans in a couple of nearby communities to give official recognition to the voyage.

Leaving New Glasgow I made the brief trip to Stellarton, Canada's first coal mining district. During his visit, Pohl had met and talked with a cross-section of people ranging from coal miners to company executives and local historians and, having done his own historical and field research, came to the same conclusion as Hobbs. They estimated it was a four-day march through untracked forest to the natural harbour at Guysborough. After closely examining its configuration and surroundings, Pohl found that it was the most likely "excellent harbour" mentioned in the Zeno Narrative. But with other matters tying up my time, another year would pass before I could make a trip up the winding eastern shore with its many jutting points and retreating coves, to Canso and then to Guysborough.

Historical records show that Nicolas Denys, who had arrived in La Have with Isaac de Razilly, the newly appointed French governor of Acadia, set up a fishing port at Canso in 1636. He also built a trading fort at Guysborough and begun a small settlement on the bluff just inside the harbour mouth. Denys, author of *A Geographical and Historical Description of the Coasts of North America*, had praised this harbour as "A fine harbour ... formed... by means of a dike of gravel 600 feet in length ... the entrance ... is a pistol shot wide, and makes inside a sort of basin ... the entrance thereto is very easy. A ship of 100 tons can enter there easily and remain always afloat. The land is very good ... Higher up there are very fine trees." Other early written accounts also testify to the natural harbour protected from the open sea and to the fertility of the land in the vicinity, the beauty of the surrounding countryside and the abundance of fish found in the local rivers.

Following Denys' departure after years of constant harassment by his own jealous countrymen, yet another French settlement was established at Guysborough Harbour by a merchant consortium out of La Rochelle. In spite of attacks by pirates and naval forces from New England, the settlement of St. Louis, so named after the reigning French monarch Louis XIV, expanded at Guysborough to the extent that by 1687 it was reported to have about 150 residents. However, the tide of history which, with the recapture of Port Royal in southwest Nova Scotia in 1710, had swept the British back in control of the province, brought an end to the official French presence. By the latter half of the century, large tracts of land in the area had been allocated to a number of British sympathisers from the rebelling American colonies and the first Protestant settlers moved in.

On a crisp, clear fall morning in 1997 I stood alone on the bluff that overlooked the mist-enveloped waters and terrain on which much of Nova Scotia's fishing history had been played out. Satisfied that the surrounding geography and natural features of Guysborough Harbour were those described in the Zeno document I wondered, as I looked out into the expanse of Chedabucto Bay, if some 600 years earlier a lone Mi'kmaq might have also stood here and seen what appeared to be an island floating towards him from the horizon, for that was the description given by the Indians to the first masted sailing ships they saw. I wondered also, if the first ship sighted from that grassy Guysborough hill had been Sinclair's of 1398, and if the arrival of the Scottish knight had given rise to the intriguing and mysterious Mi'kmaq legend of Glooscap.

7

The Glooscap Trail

"Long ago, in the misty beginning of time, when the Indian alone inhabited the country, Glooscap, god-man, warrior and leader appeared among them. From whence he came, no one knows, but it is said that he travelled from a great land in the east."

So begins one of the many accounts of the arrival of the great cultural hero of the northeastern coastal Indian tribes, the Algonquian Nation, to which the Mi'kmaq of Nova Scotia belong. The Mi'kmaq first knew themselves as Lnu'k, the People, who came to the northeast from somewhere further south. Their friendly greeting of "Ni'kmaq" (my kinsmen) was interpreted and adopted by Europeans as their tribal name. Like other North American Indian tribes, they have their own history. This oral history includes numerous stories about Glooscap, a larger than life man with magical powers. Some of the stories describe the origins of the earth and man's role in nature and are part of an indigenous sacred mythology on which native spirituality is based.

Known as Glooscap among the Mi'kmaq, he went under the name of Gluskap or Kuloscap among the Penobscot Indians in present-day Maine and Massachusetts, and stories about his good nature and remarkable accomplishments exist in native lore found over a wide area. Glooscap was always portrayed as a benevolent being who possessed magical or supernatural powers. He helped improve the lives of the people by teaching them how to hunt and fish with nets and how to gather plants and herbs for food and medicine. In the oral history of the North American aboriginal peoples, Glooscap stands out as one of the great mythological figures, a spiritual archetype. Like the other divinely imbued figures such as Quetzalcoatl of the Maya and Deganawidah and Hiawatha who helped bring peace between warring tribes by creating the Iroquois Confederacy further west, Glooscap was a teacher of peaceful ways.

Glooscap travelled around the region. He went from the Pictou area to the Bay of Fundy, and was said to make his home on Cape Blomidon, keeping a herb garden near Advocate Harbour, on the opposite side of the bay. He is reported to have been able to speak several languages and to have been adept in all the arts. From him the Mi'kmaq learned not only how to fish and cultivate the land, but also how to read and travel by reading the stars. He was a good teacher, a wise leader and a fearless warrior. But Glooscap did not stay indefinitely among the Algonquians. When he had finished his work among them he departed with a promise that he would return one day— "when you feel the ground tremble, then know it is I."

One version of the final departure of Glooscap from Nova Scotia tells that after holding a great feast for all the people at his hilltop home on Cape Blomidon he addressed them as follows: "The time has come for me to leave you. I travel to a great land far away and many will want to follow me ... Those who live a kind and just life will one day find me and live forever beside me. Those who rob and kill, who refuse to plant their crops and feed their young will always live alone."

Then, after stepping into his stone canoe, he disappeared out into the bay never to be seen in Nova Scotia again. Where he went to no one seemed to know, although one story said that he travelled to the mouth of a great river which he entered.

After reading some of the Nova Scotia Glooscap legends, first written down by the Father Silas Rand and others recorded in New England by Charles G. Leland, I assumed Glooscap belonged to the realm of native spiritual mythology and was possibly a teacher sent by the Great Spirit to the tribes of northeastern North America. He was reputed to perform such feats as moving boulders and transforming animals into rocks. He could be compared to one of the pantheon of Greek or Viking gods and yet his teachings were similar in many respects to those of Christianity. At the same time, I did not rule out the possibility that Glooscap had been a flesh and blood figure who had arrived in northeastern North America in the distant past from another part of the globe. Like others, I suspected that some of the tales about Glooscap told to people such as Rand and Leland had already been affected by cultural influences brought to North America by Europeans. Therefore, the various stories which have him in North America both prior to and after the arrival of the first European settlers suggests that he could have been anyone from an ancient Atlantean to an Elizabethan. However, the determined Pohl discovered that there were some interesting similarities between the Glooscap legend and the Sinclair saga and the characters of both men.

After returning back across the Atlantic, Antonio Zeno wrote a final letter to his older brother Carlo, in Venice, telling him that he would soon return home with the additional details of the voyage to the New World and the discoveries made there which he had already written down in a book. He also indicated that he had written a full account of the northern adventures of his dead brother Nicolo and of the life and exploits of the man he served, "a prince as worthy of immortal memory as any that ever lived for his great bravery and remarkable goodness."

Unfortunately these writings of Antonio's were either lost or destroyed and although he did return to Venice, all we know from the Venetian publication of Sinclair's exploits is Antonio's comment that after establishing a settlement in the harbour of his newly discovered country, he "explored the whole of it with great diligence."

Historical records indicating that Henry Sinclair was mysteriously absent from the public scene in Orkney, Norway, and Scotland from 1397 to his death in about 1400 suggested to Pohl and others that Sinclair could have spent up to two years on the west of the Atlantic. This further prompted Pohl and others to search for some clues that such had indeed been the case.

Hoping that he might find some oblique reference to the Sinclair visit in Mi'kmaq oral history, Pohl read that Glooscap was said to have come across the ocean from the east and to have arrived among the Mi'kmaq at Pictou. He then reasoned that Sinclair would also likely have travelled to Pictou to meet with the natives, the "great multitudes of people, half wild," whom his men had earlier seen in the area. Pursuing this line of research, Pohl uncovered up to 17 similarities between the two men. Although somewhat selective and seemingly ignorant of a wealth of other facts, they were enough to convince Pohl that Glooscap and Sinclair were in fact one and the same person.

While writing a book about the life of Sir William Alexander and the first English settlement at Charlesfort, I lived for a short time in Parker's Cove, on the eastern side of the Bay of Fundy. I soon came to admire both Cape Blomidon and Cape d'Or, from a distance, as strategic landmarks that inevitably put me in mind of the Glooscap legends since both landmarks are associated with him. Occasional visits to Halifax took me along the tourist route known as the Glooscap Trail and during these trips I often passed away the miles mulling over the theory that Glooscap had been none other than Prince Henry Sinclair. After returning back home to the south shore I had the opportunity to ask Vaughan Doucette, a Mi'kmaq from Eskasoni, Cape Breton, what he thought of the Glooscap legend. He surprised me by saying that there could have been several Glooscap

Mi'kmaq historian Peter Christmas points to a sacred symbol known to the Mi'kmaq, which is on the oldest Masonic document in Scotland, the Kirkwall Scroll.

figures spread out over a period of time, each one possessing unusual and praiseworthy abilities. On the face of it, Vaughan's answer was the most reasonable I had yet come across. It also reinforced the notion that a visit from Prince Henry Sinclair might have have been woven into the folk memories of the Mi'kmaq. And yet another confirmation of sorts came from the Nova Scotia Mi'kmaq themselves. The Sinclair conference held in Orkney, in August 1997, had among its invited speakers three prominent Mi'kmaq representatives—Peter Christmas, historian and educator, Don Julien, executive director of the Confederacy of Mainland Mi'kmaq, and Kerry Prosper, chief of the Afton band near Guysborough. At a reception hosted by the Town of Kirkwall, there were the customary words of welcome and exchange of gifts. Peter and Don presented the town with the Mi'kmaq nation's flag which displays a horizontal red cross against a white background with the images of a red star and crescent moon on either side of the cross's short arm. On seeing the flag, a Templar historian attending the reception immediately commented on the fact that this flag was practically a mirror image of a flag sometimes used by the Templar fleets during the crusades. Niven Sinclair, the London businessman who promoted the conference, pointed out in his presentation the similarity between the two flags was either a remarkable coincidence or evidence of contact

Mi'kmaq representative Donald Julien presents the map of Mi'kmaki to Niven Sinclair during the Sinclair Symposium.

between Europe and North America involving the Templars. Given that some Mi'kmaq hieroglyphics have been found to be strikingly similar to those of ancient Egypt, strongly suggesting that there was transatlantic cultural diffusion going back thousands of years, the possibility that a Templar symbol reached the shores of North America in the fourteenth century becomes highly credible. Of course, to most of the attendees this was taken as further circumstantial evidence that Prince Henry Sinclair, whose family had known connections to the Templars, had made it to Nova Scotia.

Pohl had surmised that since the Sinclair expedition contained "a considerable number of vessels and men," it must have carried carpenters, shipwrights, masons, metal workers and even clergy, in addition to the necessary contingents of sailors and soldiers. Zeno had written that even after the losses incurred during the stormy crossing, Sinclair was still able to send one hundred men on a reconnaissance mission into the interior of the newly discovered land. An equal number would very likely have been retained to defend his position at Guysborough in case of a sudden attack. So, given these numbers, the chances are that Sinclair would have had enough skilled men, sailors and tradesmen willing to remain with him in Nova Scotia to make further exploration by land and water possible, enough skilled men

Fleet battle flag of the Knights Templar (red on white background).

to build and sail a sea-going vessel that could take them back across the Atlantic to their island home should a relief vessel not arrive the following summer.

Pohl connected the reference in the Zeno document to Sinclair having "explored the whole of the country with great diligence" with the accounts of Glooscap's travels across the province to the Bay of Fundy area, where he established a winter home. The story of Glooscap's final departure from the bay in his stone canoe was therefore, according to Pohl, an account of Sinclair's farewell to Nova Scotia.

In the various accounts of Glooscap's departure we are told that when the time came for him to leave Nova Scotia, he prepared a great feast "by the shore of the great Lake Minas." He talked for a long time to those who had obeyed his commands. He smoked his last pipe with them and gave them good advice. He spoke of his going away, but of the land he was going to he would say nothing. He promised that some day he would come again among them. His great stone canoe waited for him in the bay and Glooscap said to those who had gathered to bid him farewell, "It is now sunset and I must leave you." Many of those who had admired him and followed his teachings begged him to allow them to go with him. He told them that it was not possible for him to take them on the great journey he was

Grand Council flag of the Mi'kmaq Nation (red on white background).

about to make. Then, just at the turn of the tide, as the sun set behind the distant hills, he embarked in the stone canoe and sailed out to the sea with the ebbing tide, singing as he went a strange, sad song. The people and even the wild animals were said to have stared after him until, in the deepening twilight, he was out of sight. So Glooscap sailed away to the hunting grounds of his forefathers.

The tales of Glooscap's departure from Nova Scotia also describe the deep sadness and silence that fell upon the people and the land after he had left. There is also a poetic and poignant description of the hope felt by the M'ikmaq that this benevolent individual would one day return to create a Golden Age, when men and animals would once again live together in peace.

However, the Glooscap legend also intimates that he did not immediately return back across the Atlantic whence he was said to have originally come. In one account of his departure we are told that on leaving the bay he sailed west. Such a course would have brought him along the New England coast to an area where further evidence of Sinclair's presence has been found.

8

New England Evidence

In recent years the carved image of a medieval knight found on an outcropping of rock in Massachusetts and a round tower of unknown origin in Rhode Island have been hailed as proof positive of the Sinclair landing and its connection to the order of the Knights Templar. It was found on an exposed area of rock on the side of Prospect Hill, about one kilometre from the centre of the town of Westford, Massachusetts. However, the discovery remained the subject of much controversy and speculation. Markings on the layered rock were first noticed over a hundred years ago by Rev. Edwin Hodgman who wrote "rude outlines of the human face have been traced upon it." It was assumed at the time to have been the work of local Indians. During the 1940s two local researchers examined and photographed the rock face and while doing so they made the additional discovery of what looked like the outline of a large sword in the centre of the exposed rock. Believing that it was of Norse origin and evidence of the arrival of Vikings in the area around 1000 AD reference was made to this discovery by W.B. Goodwin in his book *The Ruins of Greater Ireland in New England*.

Then T.C. Lethbridge, keeper of the Anglo-Saxon Antiquities at the Museum of Archaeology, Cambridge University,

England, on seeing a photograph of the stone carving of the sword in Goodwin's book succeeded in identifying it by the design of its hilt as belonging to the thirteenth or fourteenth century. Excited by the Westford discovery Lethbridge wrote, "This sword carved on a rock can hardly be anything but a Mediaeval sword. The whole hilt looks about A.D. 1200-1300. The significance of this is considerable. I do not see how this particular form of sword could be anything but European and pre-Columbian ... I think this sword carving is one of the most important things ever found in America."

A clearer photograph of the sword carving was later shown to an expert at the Royal Armoury in the Tower of London who confirmed Lethbridge's conclusions by officially identifying the sword as being of fourteenth century origin. In design it was found to conform to a Scottish claymore.

The fact that the carving showed the sword broken just below the hilt suggested to Lethbridge and others that it was intended as part of a memorial to a dead knight, the custom in medieval times being to break the sword of a distinguished knight on his death and bury it with his body. The whole matter then got tossed about in a sea of doubt and dispute among archaeologists and historians for about ten years until Frank Glynn, President of the Connecticut Archaeological Society, cleared the entire rock face of moss and sod and to his pleasant surprise discovered that a series of connecting punch holes extending from almost the top of the rock face formed the badly eroded image of the outline of a man. About two metres tall, his face was enclosed in what appeared to be a metal helmet and in addition to the sword, the blades of which was over one metre long and extended down the centre of the man's body, there was also the outline of a shield.

Lethbridge was the first to suggest that what Glynn had uncovered was an effigy of a medieval knight in helmet and mail dress, punch holed onto the rock by an armourer. Similar stone images had been found in northern Scotland. Since the evidence discounted any connection to the Vikings,

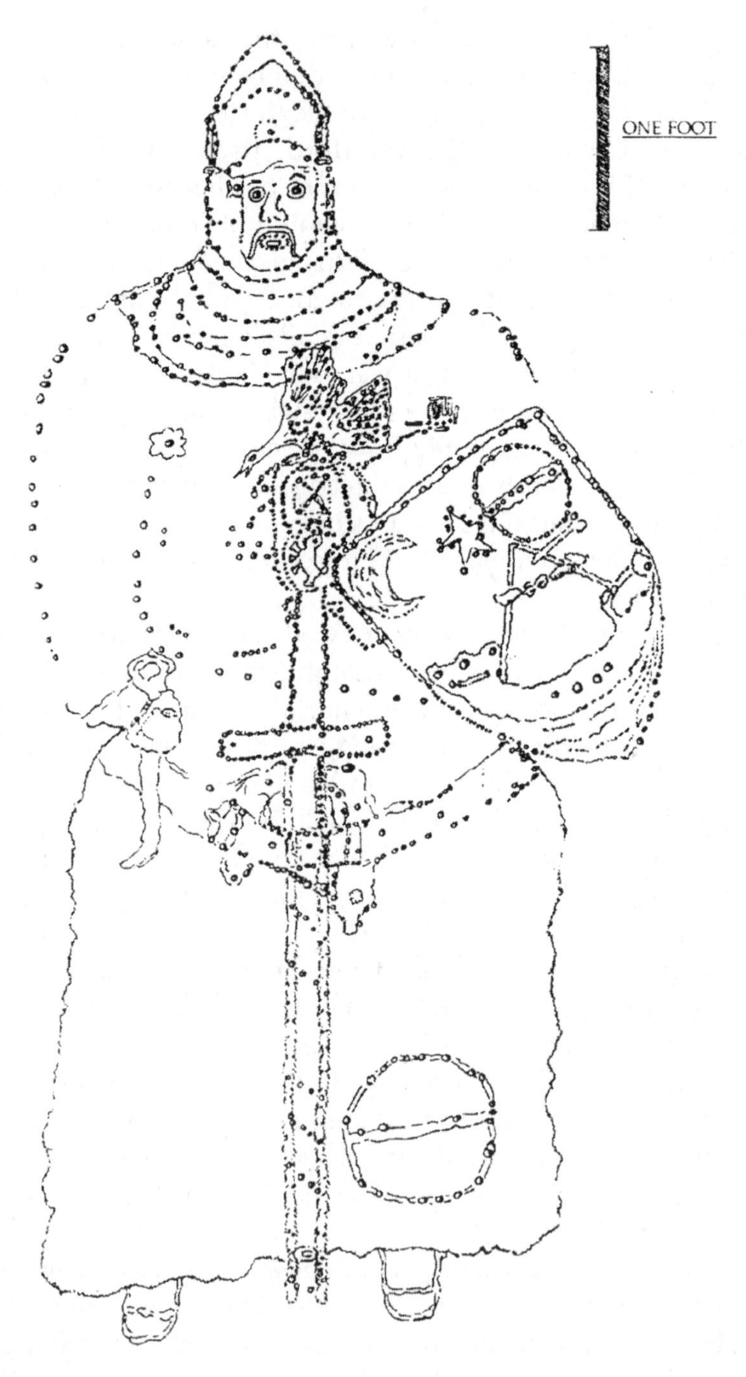

Westford Knight drawing by Frank Glynn showing the distinguishing features of a fourteenth-century knight, copied by Gertrude Johnson.

Lethbridge suggested that the effigy might be a memorial to a member of the Sinclair expedition who had died while ashore in Massachusetts.

By chalk lining the image, Glynn was able to distinguish that the shield bore a coat of arms and was even able to describe a few of the armorial bearings on it. These included a star and a circular brooch above a ship somewhat similar in outline to the Orkney galley on the Sinclair shield. In 1967 Sir Ian Moncreiffe, author of *The Highland Clan,* an authority on Scottish clan heraldry, was consulted about the images on the shield. He immediately identified these as part of the coat of arms of the Gunn Clan of Northern Scotland. The Gunns of Caithness and Orkney were also of Norse descent and were related to and were associates of the more powerful Sinclairs. Referring to the possible connection of the stone effigy of the knight to the Sinclair voyage, Moncreiffe wrote,

> There is nothing remarkable in the idea that the Earl of Orkney, a Scotsman but the premier noble of Norway, sailed to America in the fourteenth century; for the Norseman had certainly been crossing the Atlantic since at least four centuries before and the great Scandinavian houses were all inter-related. Henry Sinclair was also related to the Gunns who were at that time, the next most important family on the Pentland Firth, after the Sinclairs themselves. So the discovery at Westford of what is apparently an effigy of a fourteenth-century knight in bascinet, camail and surcoat, with a heater shaped shield bearing devices of a Norse Scottish character, such as might have been expected of a knight in Earl Henry Sinclair's entourage, and a pommelled sword of the period, is hardly likely to be a coincidence. I rather think that the mighty earl stayed a while, possibly wintered, in Massachusetts.

The logical conclusion drawn from all of this by American proponents of the Sinclair voyage was that a leading

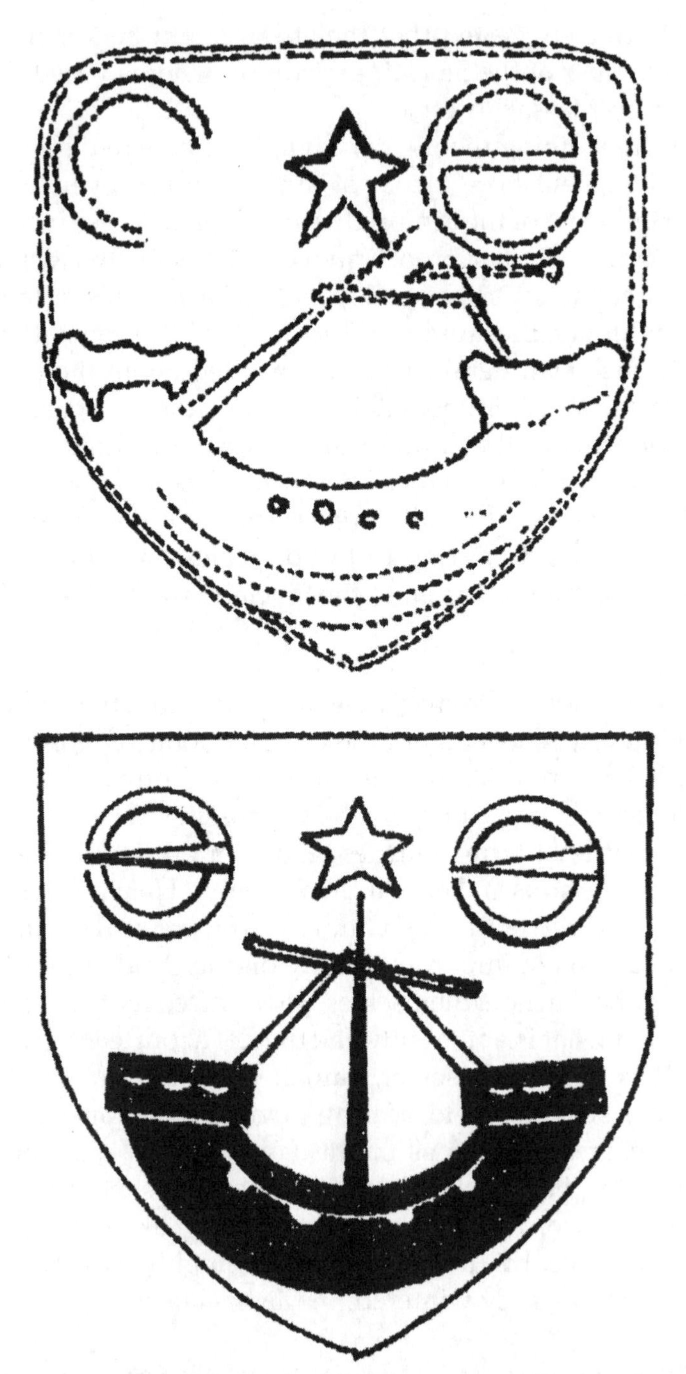

Markings on the shield of the Westford Knight (top). The shield of the Gunn Clan, closely associated with the Sinclairs (bottom).

member of the Gunn Clan had accompanied Henry Sinclair on the transatlantic expedition that had also landed in Massachusetts. For them the Westford memorial to a dead Gunn knight took on the status of a significant archaeological discovery.

While touring northern Scotland with members of the 1998 Sinclair Symposium I had the opportunity to visit the Clan Gunn Heritage Centre and Museum in a refurbished eighteenth-century church at Latheron, on the bleak and beautiful northeastern shoreline south of John O'Groats. Situated on high ground, it offers a magnificent view of the coastline and the east coast fishing grounds. Standing in the graveyard outside the building, I was struck by how similar the surrounding landscape was to that found along the Cabot Trail in Cape Breton. Inside this small museum in a remote part of Scotland, I came face to face with a replica of the drawing of the Westford knight whom the Clan Gunn has unhesitatingly and proudly claimed as one of their own. In fact, the brochure available to visitors to the museum made much of the possibility that a Gunn had been part of a Scottish expedition that made it all the way to North America almost 100 years before Christopher Columbus.

This was not the only evidence found in the Westford area suggestive of a voyage to the New World. When Frederick J. Pohl heard about the carving he visited the site to do some investigating of his own. During one of his visits he and Frank Glynn were shown a large stone on which there was a carving of a fourteenth-century ship. It had a single mast, a sail and also a number of oars. Along with the later discovery of a petroglyph, a similar ship and a cross, on a rock ledge in Machias Bay, Maine, this has been hailed as further evidence of the Sinclair voyage. The Westford stone, which had been found by a local farmer and is presently housed in the town library, also bore the distinctive carving of an arrow with a feathered tail and the number 184 to one side of the ship. Assuming that the 184 referred to a measurement, Glynn then found the outlines

Crest of the Gunn Clan at the Gunn Heritage Centre, Scotland

Scottish Medieval sword at the Gunn Heritage Centre, Scotland

of what might have once been three small stone buildings within the radius of 184 paces from the spot where the stone was found. Although no archaeological excavation was carried out in the area, these discoveries gave rise to the opinion that the site of a Sinclair encampment in New England had been found. This was reinforced when examination of

Nova Scotia and American delegates attending the Sinclair Symposium in Kirkwall, Orkney

the image of the knight by two geologists resulted in the punch holes being dated as approximately 700 years old.

The increasingly eroded image of the knight was made clearer just a few years ago when Marianna Lines, of England, did a cloth rubbing of the stone using vegetable and flower dyes. Her unique and effective method of bringing out images from carved stone has been well publicised and praised, especially for the artistic and distinctive way it both highlights minute detail and presents the total image. The rubbing, carried out with the assistance of James P. Whittall, an archaeology researcher with the Early Sites Research Society of Rowley, Massachusetts, dramatically highlighted the outline of the medieval knight in full dress. Details of the broken sword and the Gunn shield also became much more obvious. Quite understandably the enhanced image was hailed as visible evidence, literally written in stone, of the Sinclair voyage.

The summit of Prospect Hill is over 150 metres above sea level and is the highest point between the Atlantic coast and the White Mountains. On a clear day it offers a commanding view of the surrounding countryside. It has been calculated that if Sinclair and his men had sailed out of the Bay of Fundy on the strong outgoing tidal rip, a prevailing north-east wind could have taken their fixed-sail craft within sight of the coast of Massachusetts. It is known that trees on the top of Prospect Hill were visible from the ocean up to the close of the last century so it is very likely that an early explorer would have sighted the hill and headed for it. Once on shore an explorer in unknown territory would have made his way to the summit of the highest hill in order to get a clear picture of the lie of the land, just as Pohl had suggested Sinclair did after first landing in Guysborough Harbour, in Nova Scotia. A 40-kilometre hike, not a great distance in those days, along an Indian path from the Boston Harbour area would have brought Sinclair and his men to the summit of Prospect Hill. Alternatively, as Norman Biggart of the New England Antiquities Research Association has suggested, they could have travelled by boat up the Merrimack River, a trip that would also have involved some portages, as far as the Stony Brook. From there it was a mere 16 kilometres to the top of Prospect Hill.

Since the Glooscap legend was also part of native oral history in Massachusetts, Pohl considered it quite possible that it was influenced by Sinclair's arrival on the coast and travels inland. Allowing for the fact that Sinclair could have spent two winters in the New World and assuming that he had spent the first in the Bay of Fundy region of Nova Scotia, it was to be expected that further evidence of his stay in New England must exist. Supporters of this theory, of which there now are many, suggested that the mysterious Newport Tower in Rhode Island more than adequately fits the bill.

Previously hailed as evidence of pre-colonial European settlement in North America, the Newport Tower is a two-storey round tower of rubblework construction with dis-

Newport Tower, Rhode Island

tinctive medieval architectural features. Its very obvious Norman characteristics stands out on a strategic, sheltered mainland location west of Buzzard's Bay and Martha's Vineyard. This is a region where it is conjectured that Europeans intermingled with the native population resulting in Indians with distinctive non-Indian features, as noted by some sixteenth-century explorers. There is even official documentary evidence suggesting that a "rownd stone towre" existed in the area prior to the founding of Newport in 1639. Notations on a William Wood map of the region published in 1635 suggest that the remains of a pre-colonial settlement, referred to on the map as Old Plymouth, existed on the site of present-day Newport.

Although colonial documents, the 1677 will of Governor Benedict Arnold in particular, refer to the structure as a windmill and although it was used as one for a while, there is no evidence whatsoever that it was originally built for that purpose. In fact its origin is totally unknown. Certain characteristics of the structure, such as the distinctively fourteenth-century first floor fireplace and windows, reachable only by ladder from the outside, overwhelmingly suggest that it was built for another purpose. One theory was that it was originally constructed as a combination of fortification and lighthouse since the windows afforded a clear view of the water approaches to the area. However, more recent detailed examination of the structure, due to the heightened interest in the Sinclair voyage, produced evidence which indicates that it incorporates medieval architectural features commonly used by the Knights Templar, with whom the Sinclairs of Rosslyn have been long associated.

At the Sinclair Symposium, I listened to James P. Whittall of the Massachusetts-based Early Sites Research Centre make a convincing case for claiming that the tower, which stands on eight equal-sized pillars and contains Romanesque arches, was constructed in the sacred architectural style inspired by the Church of the Holy Sepulchre and the Dome of the Rock in Jerusalem, a style brought back to Europe by the Crusaders. Whittall spent six years researching the Newport Tower, and he pointed out that lines joining the bases of the ground floor arches within its circular base form a pattern which is strikingly similar to the distinctive eight-pointed Templar cross, the same as was later adopted by the Knights of Malta and in use to this day. In addition, each of the eight pillars supporting the tower, with its east-facing alignment, are placed on cardinal points in keeping with a key principle of sacred geometry used in Templar architecture. There was also the possibility, according to Whittall, that those who designed and built the Rhode Island round tower might have been influenced by the round churches of Scandinavia. The remains of a church of similar style also exists in Orphir, Orkney. Built during the twelfth century, it was similar to Templar churches constructed in Paris and in Cambridge, England. It is therefore not inconceivable that the Newport Tower served a number of functions, one of which was that of a church.

The tower contains features similar to those found in buildings constructed in the northern Scottish islands between the twelfth and fifteenth centuries. However, a number of these features, such as the fireplace with its double flues and tool marks found on worked stones, seem particular to the fourteenth century. Perhaps the most convincing fact favouring not only a pre-Columbian origin for the Newport Tower but its possible connection to the Sinclair voyage is that the unit of measurement utilised in its construction is the Scottish ell, which is derived from an earlier and larger Norse measurement. The single and double-splayed windows and the built-in stone recesses in the walls are comparable to many found in medieval buildings in Europe, in the Scottish islands and even in the bishop's palace in Kirkwall, Orkney. They are seldom found in colonial architecture. In addition, the arches and lintel design are also found in Orkney and Shetland and again in the Scandinavian round churches which were built before 1400 AD. Also, the distinctive triangular keystone above the arches, although found in buildings in Greenland and Ireland, is also common to Orkney and northern Scotland.

Originally covered in white plaster, similar to its still intact counterparts in Scandinavia, it has been described by a former director of the National Museum of Denmark as definitely medieval in origin.

Given that the Newport Tower has no architectural parallel whatsoever in colonial New England, that it was constructed in keeping with the precise geometric principles of Templar sacred architecture and that it contains a multitude of features comparable with those still existing in pre-fifteenth-century buildings in Orkney and northern Scotland, it is hard to dismiss the evidence in favour of a

Sinclair settlement in New England and of a Templar influence as well.

Remote as such a possibility seemed to me at first, after spending some time exploring the origin, history and the heroic and romantic lore surrounding the Jerusalem based, and much-travelled Templar knights, their arrival on the shores of North America in the late fourteenth century did not appear at all improbable.

9

Knights of the Temple

The first Crusade was launched by Pope Urban II in 1095 in Clermont, central France, in a bid to win back Jerusalem and the other sites associated with early Christendom from centuries of Moslem control. It was an opportunity for the Church to try to direct the socially disruptive forces of the war lords of western and central Europe in a united act of military chivalry against a common enemy, the Seljuk Turks. Hordes of these fierce horseback warriors had been pouring out of Eurasia and overrunning the Christian Byzantine Empire in the east. There was great concern that the ancient capital of Constantinople, which had been founded by Constantine the Great in 330, would be attacked. The Byzantine emperor, Alexius I, made an urgent appeal to the Pope for military aid. He could have had no idea that his appeal would give rise to a holy war involving most of the medieval world and which would have long-lasting consequences on the histories of both Europe and the Middle East.

Urban II seized the opportunity to stir up a religious frenzy that sent a number of the nobles of Europe marching east with their various armies. Their purpose, as preached by the Pope and the Catholic clergy, was no longer just to help their fellow Christians in beleaguered Byzantium but to proceed onwards into Syria and Palestine and retake Jerusalem, the original spiritual centre of the Christian world.

Unfortunately, although the Crusade was joined by numerous courageous men with sincere religious convictions, it was used by some to harass and kill many European 'non-believers,' mostly Jews, along the way. Long before they reached the Holy Land, many Crusaders, both French and German, behaved like buccaneers, plundering and pillaging, spilling blood and reaping the easily acquired spoils of an army on the move. It was not surprising, therefore, given the religious fanaticism that fuelled the Crusade and the harsh, death-dealing realities of the long journey across Europe and Asia Minor, that Crusaders indulged in an orgy of killing when Jerusalem was eventually captured some four years later. Their merciless execution of its inhabitants took no account of the moderate treatment meted out by Moslem rulers to Christian pilgrims who, for centuries, had been allowed to visit the Holy Land and pray at Christian shrines. In advocating that one could storm the gates of heaven with sword in hand, a hand covered with the blood of an unbelieving enemy, Rome had unwittingly unleashed a holocaust as horrific as anything the world has seen since. It caused a deep rift between Christians, Jews and Moslems and set a precedent for a Moslem holy war, the Jihad, and the death and destruction caused by militant religious fanatics of every persuasion that continues to this day.

The horrific overthrow of Jerusalem was unleashed by inflammatory preaching. The capture of Jerusalem had been made a sacred cause by a papal decree which also exonerated the participants from all past sins and condoned the killing of opponents of the Church. It had been the Byzantine emperor's repeated pleas for help to protect his kingdom that had given rise to the Crusade in the first place and several of its leaders had taken sworn oaths in Constantinople that if they were victorious they would restore to him his rightful eastern possessions including Jerusalem. However such matters were quickly forgotten by the triumphant

European lords who, at Rome's insistence, banished the Greek Christian Church from the city and made Jerusalem the centre of a new Latin kingdom in the East.

Most of the Crusaders and their followers returned to Europe but those who remained ruled a number of principalities stretching from the borders of Egypt to Persia. They voted to elect from among their leaders a man who would literally be the ruling monarch of the new kingdom of Jerusalem. Once elected, Godfrey de Bouillon refused the crown on the basis that he would not wear a worldly crown in the self-same city where Jesus had worn a crown of thorns. However, he accepted the administrative position and took the title of Defender of the Holy Sepulchre. Within a year of his death his brother Baldwin displayed no such scruples and was happy to participate in a coronation and to reign as the first European-born king of Jerusalem.

Nevertheless, the holy war that brought the new kingdom into being was by no means over. It continued for another two hundred years, engaging the population of the whole of Europe in several more Crusades against a determined Moslem resistance and Mongol invaders until, divided and weakened, western forces collapsed under the relentlessly savage attacks of skilled Mamluk warriors. They were finally forced to withdraw, not only from Jerusalem, but from the entire Middle East. Almost the last to leave was a group of warrior monks who came to the Holy City as a Christian chivalrous assemblage a few years after the founding of the eastern Latin kingdom, which they named Outremer, the land across the sea. These knight mystics had gained a unique place in the annals of this period. Their reputation for both devout living and heroic deeds, as well as their enormous power and mysterious ways, had made them, even in their own lifetime, the subject of legend. Because of their connection to the Sinclairs and their alleged involvement in the 1398 voyage, a knowledge of their history became essential to my research.

The legend begins around 1100. To many religious-minded people in France and elsewhere in Europe, the return of Jerusalem and the other sacred sites in the Middle East to Christian control seemed to signal that a new spiritual era was indeed dawning at this time. Arriving in Jerusalem around 1118, ready to play its part in bringing it to light, was a small disciplined band of French knights who were the founding members of the Poor Fellow Soldiers of Christ and the Temple of Solomon, more commonly called the Knights Templar. Like the Knights Hospitallers of St. John of Jerusalem, they expressed the desire to serve the needs of pilgrims to the Holy Land.

There is still some doubt as to the exact year they formed their brotherhood. A written history by Guillaume de Tyre begun in 1180 gives some information about their origin and achievements and acknowledges and chronicles their contribution to the Christian cause in the Middle East. However, mystery, mysticism and romantic tales are also very much part of the lore that survives about the Templars. In particular, their interests during the early years inside the historic, sacred Temple Mount in Jerusalem, the extent of their interaction with the Moslem and Jewish communities and their activities following their persecution in France two centuries later have remained for the most part historical enigmas. Adding to the air of mystique surrounding this close-knit brotherhood, who performed the demanding balancing act of following strict monastic rule while being prepared to fight as fiercely as the fanatical Assassins of the Saracens, was the belief that they had managed to find a priceless treasure during their years of sojourn in Jerusalem. Rumoured to be everything from the fabulous treasure of Solomon to secret texts of early Christendom to priceless religious relics, and even the Holy Grail itself, all that remains is speculation.

Supporting the hypothesis however, is the fact that the contents of one of the Essene scrolls discovered in caves beside the Dead Sea earlier this century indicated that a treasure containing unspecified valuables, sacred vessels and texts was hidden beneath the Temple. This could well be what the Templars discovered, assuming they carried out excavation work beneath their Jerusalem headquarters, as has been suspected.

What is certain is that their moral authority, military ability and accumulated wealth gave them unparalleled influence with popes and princes and gained them the unquestioned respect and adulation of the general populace of Europe. At the same time, the mystery, secrecy, power, and prestige associated with the Templars led to suspicion, fear and resentment. Apart from being perceived as the selfish possessors of almost inexhaustible wealth, exercising too much influence as power brokers between East and West and between Church and State, they were also suspected of associating closely with the "infidel"—the Moselms—and of having been contaminated by dealings with heathen sects. Some members of the other military and monastic institutions and the regular clergy resented the unique position the Templars enjoyed, for they answered to none but the Pope himself.

Regardless of the uncertainty about the exact year of their origin, most historical accounts agree that by 1118, with Jerusalem and the region safe in Christian hands, the French knight Hugh de Payens, a vassal of the Count of Champagne in northeastern France, and eight associate members of the French nobility had established the order of the Knights Templar, so called because their headquarters were situated in the Al Aqsa mosque on the mount where tradition had it the magnificent Temple of Solomon had once stood. This had been the site of the Holy of Holies, making it the most sacred place to Jews, as it was to Moslems because Mohammed had been blessed with a heavenly vision on the same spot. For Christians, this site was dramatically intertwined with the life of Jesus. The Templars' initial purpose, as expressed to King Baldwin who gave them part of his palace compound on the Temple Mount, was to protect and assist pilgrims to the Holy Land. Another chivalrous order, the Knights Hospitallers of St. John of Jerusalem, had already established both a hostel and hospital for pilgrims in Jerusalem.

It seems quite likely that the motivation behind the creation of the Knights Templar arose out of a liberal Christian evangelical movement in parts of France that had one foot in the garden of eastern esoterica and the other firmly planted behind the walls of western monasticism. The city of Troyes, in the Champagne region, was a centre of esoteric studies and had gained a reputation as a place where alchemists and astrologers broke bread and shared a jug of wine with artists, poets and Biblical scholars. The court at Champagne supported the writing of the first of the popular Grail romances, which married metaphysics with Christian sacred symbolism in a secularised telling of the gnostic tale of the soul's abiding need and burning desire to know God. The court also fostered the emergence of the importance of the feminine in medieval thought and the cult of courtly love. Of course, it is possible de Payens and his fellow knights were merely giving chivalrous expression to the religious temper of the times, which called for acts of penitence and self sacrifice in service to a sacred cause. Whatever the source of their initial inspiration, they certainly made their presence felt in Jerusalem and quickly gained recognition for their religious zeal back home in France. However, in the oftentimes precarious and hostile environment in which they found themselves in the Middle East, their emphasis soon shifted from the role of pilgrim protectors to that of proactive militant defenders of the Holy Land and the true faith.

Bernard of Clairvaux, one of the most authoritative and respected voices of the Roman Church at the time, said of them, "They are milder than lambs and fiercer than lions. They combine the meekness of monks with the courage of knights so completely that I do not know whether to call them knights or contemplatives." He also noted that unlike knights, they did not hunt or hawk, avoided pomp and decorous trappings and despised "mimes, jugglers, storytellers and dirty songs." Bernard, who belonged to the monastic Cistercians and had also benefited from the Champagne coffers, drew up an austere religious rule for the expanding Templar order. It required them to take vows of poverty, chastity and obedience and to live communally. They were restricted as to their manner of dress, diet and behaviour. They exchanged their rich robes for white mantles symbolising their entry into a purer way of life, much like the Essenes of old and the Parfaits of the Cathars. Although abandoning the rich fare they had been used to, they were permitted to eat meat three times a week and a daily dose of wine. Abjuring idle and profane talk, they took their meals in silence while listening to uplifting readings from the scriptures. Physical contact with a woman was forbidden, as was sleeping naked or the use of any sensory stimulus. As a mark of distinction, if not convenient identity, they were required to remain unshaven. Apart from the knights, there was a large number holding the rank of sergeant. Trained in strict and unbending military discipline they were expected to fight to the death in the defence of their faith, like the Ishmaelite initiates in Egypt and the drug taking Hashishin in Syria. Such sacrifice and dedication to the Christian cause earned them the admiration and support of many in Europe so that this holy brotherhood of courageous knights was officially endowed with the status of a religious order in 1129 at a Church Council which met in Troyes.

Even though they had not participated in the victorious First Crusade, nor been founded for the purpose of military conquest, the Knights Templar developed into the major permanent western fighting force in the Middle East. Their knowledge of military strategy, the nature of the enemy and the geography of the region helped them win many a battle. Their remarkable courage, determination and discipline in the open and on the ramparts could always be relied on. They would retain that position, playing a prominent role in several of the Crusades during much of the next two hundred years. The Knights of St. John of Jerusalem followed suit as did the Teutonic Knights, who began life as a German brotherhood in the northern coastal city of Acre during the Third Crusade. Between them, these three military orders became the crack storm troopers in Outremer. However, all too often, ambitious European lords at the head of ill-prepared and poorly equipped armies ignored the seasoned advice and the leadership of the Templars or argued among themselves and suffered devastating setbacks and defeats.

As previously mentioned, the Templars' activities were by no means solely military. The order benefitted from donations of land and money from the members, as well as receiving bequests of extensive properties and numerous other gifts from many of the noble families of Europe. During their almost two hundred years of existence, the Knights Templar became administrators, not only of strategic castles and fortress towns in the Middle East, but also of a vast network of European landholdings which in turn helped finance their continued presence and activities in Outremer.

They constructed and maintained preceptories and fortifications, incorporating the design of the Church of the Holy Sepulchre and the eight-sided and arched Dome of the Rock, into their own churches. Because it exercised control over communication routes by land and sea, between the East and the West, the order was also able to profit from the rich trade carried on with the help of the Venetians and the merchants of other Italian cities. With a chain of preceptories straddling whole continents and a reputation as a well protected and wealthy international brotherhood, it was not surprising that it also developed into what was probably the first transnational financial institution, one capable of bankrolling cash-strapped kings, struggling princes and even the occasionally impoverished pope.

Being resident in the Middle East also gave the Templar Knights the opportunity to acquaint themselves with the cultural life of the countries in which they served. Peaceful interaction with both Moslems and Jews had beneficial and enriching repercussions. They became acquainted with eastern concepts in art and architecture, with new varieties of foods and clothing and more effective medicines and with the poetically evocative literature of the peoples who inhabited the hills and valleys, towns and cities of these ancient lands. The French historian Fulcher of Chartres in residence in Jerusalem observed at first hand this cross-cultural fertilisation and wrote of its effect on the transplanted Europeans: "For we who were Occidentals have now become Orientals. We who were Romans or Franks have in this land been made into Galileans or Palestinians. We who were of Reims or Chartres have now become citizens of Tyre or Antioch. We have already forgotten the places of our birth."

As members of a religious order whose daily routine, when not in combat, was akin to that of a contemplative religious community, they were certainly expected to be interested in spiritual matters. Undoubtedly some more than others sought to develop their spiritual life while living in the Holy Land, for this was the region of the world into which the ancient mystical teachings of Egypt, Persia and India flowed like tributaries of one expansive river of knowledge, irrigating and enriching the fertile Jewish religious landscape from which Christianity flowered. In an environment charged with a holy history capable of having a profound influence upon the mind and heart of the believer and surrounded by Biblical sites and scenes from the life and times of Jesus, there was always the possibility that one might encounter a heightened sense of the Divine. In a land where there was also the very real possibility of discovering sacred records or relics, any such experience could well have given those motivated and daring enough the desire to explore more deeply into the metaphysical background of Christianity, which had developed over its one thousand years of existence from a fringe Jewish sect into a highly institutionalised, politicised and hierarchical religion.

It is certainly reasonable to conclude that the impetus for any Templar spiritual adventures could have come from the discovery of sacred texts of an esoteric nature. Early this century Biblical texts were found at Nag Hamadi in Egypt and many more later at Qumran by the Dead Sea in Israel. The translation of the Dead Sea Scrolls has given us insights into the teachings and lives of the Essenes, the obscure, mystical Jewish sect that existed in relative isolation for hundreds of years prior to the birth of Christ. More important still, the contents of the scrolls have provided us with the knowledge that there were specific historic, ritualistic, and doctrinal links between the Essenes and the Nazarenes, as the early Christians were called. They were a persecuted sect in their own land, as the Essenes had been before them, and it is distinctly possible that documentation dealing with their early activities had been hidden away.

Assuming the Templars' interest in excavating and searching for precious religious relics and buried treasure had brought such documents to light, they could well have been able to translate such texts with the help of local Arab or Jewish scholars. Such a discovery could well have presented startling new evidence of Christianity's beginnings, evidence that may have been kept secret among a circle of select members within the order.

The Templars' occupation of the site of the temple that was the very centre of Jewish worship not only presented the possibility for finding informative historical material or even treasure buried beneath its foundations but it also connected them to the fascinating body of lore related to its initial construction. Containing elements of Hebraic sacred tradition, Hermetic science and Pythagorean principles of structural harmony, the building was in itself a repository of spiritual knowledge. Given the fact that a number of Masonic historians admit to an historical and philosophical link between their fraternity and the Templars, it is not surprising that the legend of the construction of King Solomon's Temple is part of the initiation rites used in modern Masonry.

During a period of research in England I spent some time in London and visited the twelfth-century Templar Church situated between the Embankment and the Strand. While crossing the Thames by Waterloo Bridge I could see the majestic dome of St Paul's Cathedral not far away and it reminded me that eastern architectural innovations passed

Templar Round Church, London

Effigy of crusader knight in the Templar Church, London

on to the European guilds by the Templars had enabled Sir Christopher Wren, a Master Mason, to grace the London skyline with a form that had its origin in the Orient. Entering the grounds of the Inns of Court, I made my way to the Temple Church, the finest surviving round church in England, built as far back as 1185 on land donated to the Templars by King Henry II. Modelled on the design of the Church of the Holy Sepulchre in Jerusalem, its slender pillared arches ascend beneath a central cupola. Within the circle of its tiled floor stone effigies of long dead knights lie gazing heavenwards like silent sentinels of a military order whose interests extended into the realms of sacred architecture, symbol and spiritual quests.

When I returned to Nova Scotia I visited another round church, not as old but equally symbolic. St George's Church on Brunswick Street in Halifax also carries within its circular design the sacred architectural heritage of the Templars, passed along the centuries through the ranks of Freemasonry. For, as Professor Atilla Arpat, formerly of the School of Architecture at the University of Istanbul, has proven by his meticulous research and pointed out in an article published by the Nova Scotia Historical Society, St George's demonstrates the system of sacred geometry long utilised by the Masonic order. Within this and other churches, Arpat repeatedly found the use of the measurement 318, which in mystical numerology represents Jesus Christ. The Masons had learned that God was the Divine Architect of an harmonious universe. They believed that in the days of Adam and Eve mankind had been given the knowledge to live and create according to the same laws of harmony. Unfortunately it had chosen otherwise and that awareness and ability was lost. But the prophets of old, the sages and seers of the East, maintained the knowledge of mankind's spiritual identity until a time when a new redeeming Adam and a new Eve would appear.

This may have been the great secret made known to the Templars. They had discovered as a result of their direct contact with the East that there is but one God abiding equally in the heart of every sincere Christian, Moslem, Hindu and Jew. This simple but startling discovery, later eloquently echoed by the English mystic poet William Blake, enabled them to realise, as had the Essenes, the Nazarenes and the Gnostics, that the real temple, the Temple of the Living God, was the sanctified human body as spoken of by Jesus when he addressed the outraged priests in Jerusalem, the same body he raised from its lifeless state three days after the crucifixion. That cosmic act opened the doors of spiritual perception so that men and women could again clearly see that they too could live upon the earth as spiritual beings.

This knowledge was only available to those trained in the ways of the spirit and who were judicious enough to use it wisely. It was also a dangerous knowledge for it flew in the face of a Church whose dogma demanded allegiance and compliance under pain of excommunication and even death. So this knowledge was safely hidden in sacred symbols and stories, and in arts and crafts. The mathematics and geometry, the tools, tales and marks of the master builders of the Gothic cathedrals in Europe became the secret sources for the transference of an esoteric Christianity that later found a home in Freemasonry.

Their long sojourn in Outremer introduced the Templars also to the chivalry, scholasticism and piety of their declared enemies. To their surprise, some battle scarred knights found themselves being treated with the utmost care and consideration by their Moslem captors. What lesson in compassion and religious tolerance could have been greater than that demonstrated by the renowned Moslem leader Saladin when he recaptured Jerusalem almost one hundred years after the Crusaders had taken it. In contrast to the butchery permitted by the leaders of the western armies, Saladin exercised enormous restraint. When his forces retook the city in 1187, the noncombatant inhabitants, regardless of their religious affiliations, were spared the sword although many ended up in captivity. There was no pillaging of property and the most sacred of Christian shrines in the city, the Church of the Holy Sepulchre, was protected and later made accessible to Christian pilgrims, thanks to a treaty arranged by the Byzantine Emperor Isaac Angelus. Of course, on the battlefield the Moslem leader gave no quarter, as was demonstrated just three months earlier at the decisive defeat of the Christian forces at the Horns of Hattin, overlooking the Sea of Galilee. Poor judgment by the Templar Master, Gerard de Ridfort, and expert Moslem military strategy led to a pitiful slaughter that culminated in the ritual beheading of all captured Templars.

For another hundred years the Templars contributed to the European military, merchant and religious presence in Outremer, which struggled vainly through successive Crusades to regain its former size and glory. But having survived a threat from the seemingly invincible Mongol forces of Genghis Khan, who butchered 80,000 of the citizens of Baghdad on his sweep westwards from the steppes of Asia and then moved on to Damascus, the eastern Latin kingdom went through a period of attrition and decline. Years of intrigue and competition among the local Christian rulers, the growing unwillingness among European princes to support Outremer's continued existence, the greed of Venetian and Genoese merchants, as well as the ascendancy of Mamluk military power throughout Palestine and Syria, set the stage for its dramatic final days. In 1291 the combined Christian forces were almost totally decimated with the loss of the coastal stronghold of Acre and other toeholds in Palestine and Syria. In an assault too overwhelming to withstand, the wheel of karma came full circle and untold numbers of Christians were slaughtered by the victorious Moslems.

The Templars, still wealthy and powerful, moved their headquarters to Cyprus hoping to re-establish themselves one day on the mainland. It was not to be. In 1307, Grand Master Jacques de Molay and 60 Templar knights visited France to discuss with Pope Clement V the possibility of a new Crusade. While staying in their palatial Paris citadel they were suddenly arrested by the soldiers of the French king.

Philippe le Bel was an autocratic monarch intent on dictatorial rule of a much expanded France. To achieve his ends he broke the power of the Church by arranging for the murder of one Pope and the election of his own choice, a Frenchman, to the Throne of Peter. He then had the Papacy moved from Rome to Avignon where he could control an obligated and terrorised pope. Always in debt, in spite of the excessive taxes he imposed on all and sundry, he owed large sums of money to the Templars as he did to the Jewish and Lombard bankers. After exiling the Jews and the Lombards, he seized the opportunity to rid himself, and the country, of the Templars.

Apart from Philippe's debt to the Templars and his interest in acquiring their wealth, he resented their power and position in French society. Like a number of other people across Europe, he was of the opinion that the Templars and the other military orders had been primarily responsible for the loss of Jerusalem and the Eastern kingdoms and had therefore outlived their purpose. He also shared the view that the Templars had enjoyed too cozy a relationship with the 'Infidel' over the years and that their faith had been contaminated and their morals corrupted by such contact. The Templars had in turn refused him admission to the order.

At dawn on the morning of Friday, October 13, 1307, in an act of audacious perfidy, all the Templars in France were arrested by order of the king, along with their Grand Master. Official records show that only about twelve Templars escaped. All Templar possessions were confiscated and all of the order's properties in France were placed under Philippe's control. He then laid a series of defamatory charges against them. He accused them of denying Christ, spitting on the crucifix, worshipping idols and practising homosexuality. To extract self-incriminating confessions, he subjected many Templars to excruciating torture. Naturally there were admissions and in spite of a Papal commission

Templar tombstone in St. Magnus Cathedral, Kirkwall Cathedral, Orkney, showing the images of the rose within the grail (top) and the sword (right).

that found very little to back up these accusations, 54 Templars, including Jacques de Molay, were burned at the stake as heretics. In 1312, Pope Clement V, under pressure from the vindictive king and still without any proof of guilt, suppressed the order, without condemning it. Whatever genuine sins the Templars may have committed, and they certainly had their share of faults, they did not deserve such an ignominious official end.

But even if the Templar order was officially down and out, many of its members were still very much alive and well. While the French king benefitted financially by its dissolution, which resulted in all Templar properties being handed over to the Hospitallers, neither his forces nor the Pope's edict could bring all the Templars in Europe to heel. Although the order no longer had the Church's blessing, few individual Templars were persecuted outside France and several found refuge and further service in other military orders. Some of those who escaped from France sailed northwards to safety in Scotland, whose king, Robert Bruce, had already been excommunicated by Rome after killing a rival in a church. Templar gravestones dating back to the period have been found in both eastern and western Scotland, as well as in Orkney. No doubt some Templars joined their brothers at their Balantrodoch Preceptory under the protection of the Sinclairs of Rosslyn. The Sinclairs eventually inherited the Templars' sacred lore, if not their wealth, and immortalised their role as loyal guardians of an ancient spiritual knowledge in Rosslyn Chapel, built on Sinclair lands, south of Edinburgh, often referred to as the Chapel of the Grail.

10

The Legend of the Grail

That some members of the Sinclair family were initiated into the Templar order, possibly at its Scottish headquarters not far from Rosslyn Castle, there can be little doubt. The evidence exists for all to see inside Rosslyn Chapel, built between 1446 and 1500 by Henry Sinclair's grandson William and described by one authority as being in some respects the most remarkable piece of architecture in Scotland. Within its buttressed walls and beneath its vaulted roof resplendent with carvings of roses and stars, one finds numerous religious allegories cut into stone. Among the array of images culled from Celtic, Nordic and Christian motifs, one can see evidence of the Templars' legendary association with the Holy Grail, representations of which exist on every side. Here too, among the carvings of flowers, fruit and foliage, are rows of Indian corn and Aloe cactus, both native to North America. Many believe that they were copied from plants brought back to Scotland by Prince Henry Sinclair.

In spite of the forced confessions in France, the Inquisition trials and the Pope's decision to disband the order, the Templars were left relatively undisturbed in Scotland and their properties remained under their own control for much longer than they did in Europe. According to Scottish Templar archivist, Robert Brydon, and Templar historian, Tim

Rosslyn Chapel, west entrance and window with Sinclair engrailed cross.

The Apprentice Pillar in Rosslyn Chapel, with carvings representing the Tree of Life rising from the entwined serpent, at the base. In the foreground is an image of the Grail, with which the chapel is closely associated.

Sinclair engrailed cross on centre altar in Rosslyn Chapel, Scotland

Wallace Murphy, the order managed to function as an underground cell of the extensive public organisation it once was. Tradition has it that a group of mounted Templar knights played a key role in King Robert Bruce's victory over the English at the historic battle of Bannockburn in 1314.

It had long been suspected that in Scotland this brotherhood of soldier monks secretly passed on their arcane knowledge and mysterious initiation rites, which were infused with Cabbalistic, Hermetic, Sufi and Gnostic influences. On viewing the numerous stone carvings that grace the roof, walls, pillars, windows and doorways of Rosslyn Chapel it was quite clear to me that Sir William Sinclair, the Grand Master Mason who designed it, had an understanding of the sacred symbols and themes so important to the Templars. And while admiring the craftsmanship that has caused this remarkable chapel to be called, as were many of the great cathedrals of Europe, a "Bible in Stone," I repeatedly came across the image of the Grail.

The presence of a carved Templar chalice or grail on the gravestone slab of Prince Henry's grandfather, an earlier Sir William Sinclair, is proof positive that the Sinclairs had aligned themselves with the Templar order many years prior to the historic voyage of 1398. The gravestone, which had been moved from a previous and smaller chapel on the grounds of the present one, is testimony to the fact that the man who died fighting the Moors in Spain was an initiated Templar knight. The eight-pointed rose contained within the grail cup and the seven-stepped base on which it stands are distinctly Templar in origin. This grail is a reminder of the cup which Jesus shared with his Apostles at their final Passover meal together, known to Christians as the Last Supper. To the Templars and many other esotericists of the time, the rose represented the covenant of divine love which Jesus pronounced — an all embracing and forgiving love. The rose's configuration within the Rosslyn grail is distinctly Templar, inspired by the eight-sided Church of the Holy Sepulchre, which was built atop the tomb that tradition says once held the body of Jesus. The seven steps at the base of the Grail signify the seven steps which led to the door of the Holy of Holies within the Temple of Solomon, but they also represent the seven stages through which the spiritual initiate had to ascend to be nourished by the contents of the Holy Grail of Christendom, the Christ-like love contained within the enlightened mind and compassionate human heart.

However, more than two centuries before this chapel was built and more than a hundred years prior to the birth of Prince Henry Sinclair, the Templars' status as protectors of this most precious relic and sacred symbol of Christendom had been publicly heralded by the great medieval *minnesinger*, the Bavarian born mystic and poet, Wolfram von Eschenbach, in the legend of the Grail quest. It would require much more than a single chapter to give a comprehensive account of its history and to do interpretative justice to the legend of the Grail. However, it is a subject that has become so interwoven with the Sinclair saga that some understanding of its origins, its place in the medieval mindset and its meaning is called for.

In the heroic literature of the Middle Ages no one subject exercised a greater fascination than the cycle of romantic stories that dealt with the quest for the Holy Grail. Taking their cue from the southern French troubadours, medieval minstrels delighted their audiences all across Europe with various accounts of the adventures of those knights who set out in search of the Grail. Although the quest is predominantly Christian in its present form, its underlying theme of the search for the sacred, for a sense of oneness with the divine, has existed since time immemorial.

Stories of this quest have invariably involved a vessel of some kind that contained magical properties or miraculous powers. Among the Celts it was a large pot or cauldron containing the food and drink of the gods, which alone would satisfy the hunger and slake the thirst of the worthy warrior. Alternatively the contents of this container were alleged to be capable of restoring life to those heroes slain in battle. One such folk tale perpetuated by the eleventh century writer Bledhericus is believed to have originated in Wales centuries earlier. Another has an Irish origin and stems from the days of the Tuatha de Danann, the mythical and semi-divine early inhabitants of Ireland. The Egyptians and the Greeks, those spinners of some of the best stories of

Gravesite of King Arthur and Queen Guinevere at Glastonbury Abbey

ancient times, told of a cosmic cup, a *krator*, from which the creator of all life dispensed to each human being individual portions of awareness, intelligence and wisdom. In the original stories of the Grail associated with England's sixth-century King Arthur and his heroic warriors, it is a magic cup, the loss of which has weakened the king and laid desolate his kingdom. In the Middle Ages it came to represent the wine cup used by Jesus at the Last Supper or the dish which held the meats served at the same meal. In fact the word grail actually derives from the Latin word *gradalis* or the old French word *graal* both of which refer to a dish or platter brought to the table at various stages of a meal in medieval times.

Emerging out of the European romantic literary movement of the twelfth and thirteenth centuries, a period that also saw the development and spread of the civilising code of chivalry, no theory of the origin of the Grail legend has behind it the weight of established historical fact. The earliest version of the symbol laden legend in the popular literature of the time was *Le Conte de Graal* or *The Story of the Grail* which was written by the French court poet Chrétien de Troyes around 1180. Crediting as his source a long lost prose version of the story, which he claimed to have been given by his patron Count Philippe of Flanders, a cultured nobleman of the period who became Regent of all France, de Troyes' recounted in poetic form the heroic legend of the Grail quest. It was clearly based on earlier tales about the adventures of England's King Arthur and his knights. Such tales had come to northern France as early as the sixth century courtesy of travelling Celtic storytellers.

De Troyes may also have been influenced by his earlier benefactor, Mary of Champagne, the educated daughter of the free-spirited and independently-minded Eleanor of Aquitaine. Both of these women have been credited with introducing to French society the cult of courtly love that sought to replace the somewhat boorish behaviour of their menfolk with a more mannered code of conduct, the same as was said to have existed in Arthur's court at Camelot. At the same time, this twelfth-century women's movement fostered affairs of the heart between men and women above the church sanctioned sexual relations based on property and privilege and it gave the feminine influence a more prominent place at the cultural if not the political table. The court at Champagne was in fact a centre of international culture and learning where artists, musicians and poets mingled with alchemists, mystics and astrologers. Influenced as he must have been by such a milieu, it was not surprising that de Troyes' telling of the tale has an untutored country bumpkin named Perceval of Wales embarking on a series of adventures which eventually takes him to Arthur's court and involves him in the quest for the Grail.

De Troyes' epic poem, although not finished before his death, contains elements of Arthurian fable, Celtic lore, Christian idealism and Eastern mysticism. It ignited the medieval imagination, was an instant success and spawned several other versions.

The Grail quest story that most notably fused eastern and western symbolism and integrated sexuality with spirituality is undoubtably *Parzival* , which was penned around 1207 by Wolfram von Eschenbach. Being a knight who had visited Jerusalem and the Holy Land during the time of the Crusades he brought the element of contemporary historical realism to the story. It was von Eschenbach who first associated both the Knights Templar and the French Gnostic Christian sect known as the Cathars with the legend. His *Templiessen*, dressed in white surcoats with red crosses similar to the garb of Templar knights, guard the Grail Temple on Monsalvasch, which is a thinly veiled reference to Montsegur, the Cathar stronghold in the Languedoc region of Southern France. Von Eschenbach asserted that the source of his version was a Jewish, Cabbalistic philosopher named Flagetanis in Toledo, Spain. Under the Moors, Toledo had been a centre of eastern esoteric influences and Flagetanis claimed that he came across the story while translating Greek writings on ancient Hermetic lore into Arabic. Intermingling his own experience in the Middle East with sacred traditions and myths of the East and cultural interests in the West von Eschenbach expanded the tale, increased its symbolic content and gave it a contemporaneous setting. In view of the Templars' growing reputation as an independent, renegade religious order and the heretical nature in which the Cathars were viewed by the Church, his version also had a topical edge.

His hero, the French equivalent of de Troyes' Perceval, is a young rustic smitten by the overwhelming desire to become a knight at King Arthur's court. In spite of his mother's protestations, he mounts an ass and rides off unconcerned by the fact that he looks more like a fool than a knight. After killing a wayward knight who had insulted

Queen Guinevere, he claims the dead knight's armour, weapons and steed. He is convinced that having proven his prowess and possessing all the exterior trappings, he has automatically entered the ranks of knighthood. Of course he is still woefully ignorant of the ways of a true knight. This lack is set right by the tutoring of an old knight who befriends him and even offers him the hand of his only daughter in marriage. However Parsival, now conscious of the chivalric code expected of him, declines and sets out to win the hand and heart of his lady by deeds of valour. Letting his horse guide him on his way, he is fortuitously led to his soul mate Condwiramur, his "guide to love" and on her account passes a requisite knightly test. His reputation now established at the court, he marries and settles down. A few years later, having fathered two sons and made his mark in the world, he decides to embark on the quest for the elusive Holy Grail. Bidding his family goodbye he again lets his trusty steed guide him to his destination.

This time he is taken to the Temple of the Grail that stands in the midst of a wasteland ruled by a wounded king. The monarch's only hope for healing lies with a knight who can respond correctly to the mystery and majesty of the Grail when it is brought before him. Unfortunately Parsival fails this test and is left to wander as a disgraced outcast made only too conscious of his shortcomings. Bitter and disillusioned he meets up with an understanding hermit who explains the true meaning of the Grail quest. Finally he is able to fulfil the destiny he has set himself. His accomplishment frees the wounded king from his suffering and restores the surrounding wasteland to fertility. In the final outcome Parsival is reunited with his wife and children and is honoured with the crown of the Grail King.

It was the thirteenth-century Burgundian poet Robert de Borron who introduced the character of Joseph of Arimathea into the story, likely drawing from apocryphal writings relating to early Christianity. He identified the Grail with the cup used by Jesus at the Last Supper and also believed to have been used to catch some of the blood flowing from his body on the cross. In de Borron's version it clearly symbolises the high and holy state of Christian love which the spiritual warrior sets out to attain.

De Borron's account of the legend, known as *Joseph d'Arimathie* and also *Le Roman d'Estorie du Graal* is full of Christian motifs. De Borron claimed that his source was a book written by early church clerics. In saying this he may have been hinting at *The Gospel of Nicodemus*, which was written during the first century AD. Central to the story is the character of Joseph of Arimathea, the wealthy tin merchant who had joined the Followers of the Way, as Jesus' disciples were sometimes called. It was this Joseph who had sought permission from the Roman governor, Pilate, to have Jesus' body removed from the cross and who provided the tomb for the burial. According to de Borron, Joseph was accused of having stolen the body of Jesus following the Resurrection and was imprisoned. While languishing in captivity he was visited by the risen Christ who presented him with the Grail and requested that he become its protector. After being freed from his long captivity around 70 AD the aging Joseph, along with some of his family and other first-century Christians in Palestine, embarked on a circuitous journey that ultimately took them to Glastonbury in southern England. Joseph then used the Grail to help establish the first Christian community in Britain. After Joseph's death, his son-in-law Bron becomes its caretaker and it ends up being buried as a treasure too precious to risk losing. But during the ensuing centuries the sacred object went missing again so that by King Arthur's time, when the country and Christianity were being threatened by hostile forces from within and without, the search for the long lost Grail had become an urgent concern of the court. But the Grail was much more than a physical object.

In *La Queste del Saint Graal*, which many Grail researchers believe was written during the early twelfth century by Cistercian monks, whose order had very close associations with the Knights Templar, the legend of the Temple of Solomon was woven into the story of the Knights of the

Round Table and their search for the elusive Grail. Having preoccupied those at King Arthur's court for some time it suddenly and miraculously appears one day to the assembled knights in a vision of spiritual splendour. Following a clap of thunder it floats in a radiant light above the heads of all present. It is covered by a transparent veil and accompanied by a most pleasing fragrance. Here the authors are at pains to identify it as something that is not of this earth. It is clearly presented as a "seeking out of the mysteries of our Lord, the divine secrets which the most high Master will disclose." Its supernatural power is such that during its presence each knight is enabled to see the others in a glorified state. When the captivating vision disappears the knights dedicate themselves to finding the hidden Grail. Among the knights who set out in search of it, encountering various challenges along the way, is the familiar figure of Perceval.

Sir Thomas Malory's *Le Morte d'Arthur* completes the medieval cycle of legends about the Grail. Written around 1485, close to the time of the completion of Rosslyn Chapel, it is similar in content to the version penned earlier by the monks at Glastonbury.

With their combination of sacred and secular symbols and motifs drawn from an eclectic mix of cultural sources the collection of Grail stories formed a substantive body of populist, esoteric literature. As part of an evolving European cultural movement it gave unprecedented emphasis to the notion that spiritual, mental and physical aspects needed to be integrated within the context and experiences of everyday life. Although this concept infused western Christianity with a new dynamism, it was viewed with some suspicion by the Church, as many discovered to their peril.

The persecution of the Cathars in the early thirteenth century renewed interest in the mystery of the Grail because of their association with it. Stories of their remarkable faith and courage spread rapidly throughout Europe and added impetus to the spiritual and cultural renaissance shortly to break out in the southern part of the continent. For the Cathars, like some other so-called heretical sects, had embodied the Gnostic philosophy inherent in the legends of the Grail. It was therefore not surprising that it was believed that the Cathars had acquired the Grail and that they had managed to send it to a secure hiding place prior to their final decimation at Montsegur in 1244. Similarly, it has been suggested that the persecution of the Templars less than one hundred years later only added to the determination of the remnant to continue to protect and perpetuate the philosophy and treasure they had discovered and which had empowered them. Certainly the nature of the sacred symbolism associated with the Templars, which was later adopted by Masonry in Scotland, points to their having been conversant with the transcendental knowledge of the ancients. Such knowledge needed to be protected and preserved for future initiates. It is therefore reasonable to assume that it had been carried to Scotland by the Templars, who in time would set their sights on creating a new Avalon and building a new Jerusalem in the comparative safety of new lands across the North Atlantic.

Of course, other interpretations of the Templars' secret and of the meaning of the Grail exist. The theory proposed in the 1982 book *Holy Blood, Holy Grail*, that the Templar treasure and the Grail both represented a blood line resulting from a marriage between Jesus and Mary Magdalene, was based according to the authors on the "crucial piece of evidence" that the word Sangraal referred to such a possibility. This involved a convenient piece of etymological conjuring by the authors, who suggested that the word in question could be rearranged to read Sang Raal meaning Royal Blood. They then suggested that this blood line continued through the Merovingian French kings to this day. This theory was in turn taken up by Michael Bradley in his book *Holy Grail Across the Atlantic*. However, it ignored the allegorical nature of the legend and was lacking in any explanation of the spiritual and psychological symbolism involved. As author Andrew Sinclair has pointed out in *The*

Sword and the Grail, no bloodline linking Jesus to the former kings of France ever existed. In 1997 a BBC investigative television programme revealed that the evidence which the authors of *Holy Blood, Holy Grail* were alleged to have discovered in France in relation to the Templar treasure was phony. A documentary entitled *The History of a Mystery* disclosed that "secret documents" which the authors came across in the *Bibliothèque nationale* in Paris, and on which, to a large extent, they based their theory that the Grail represented a royal blood line extending back to Jesus, were fabrications and had been placed there by two French conmen, Pierre Plantard and Philippe de Cherisey. They profited from the elaborate deception and led many people to erroneous conclusions. This misleading information pervades several recent publication and internet sites on the topic of the Grail.

The link between the Knights Templar and the Sinclairs of Scotland is an established historical fact. While growing up at Rosslyn, Henry Sinclair would certainly have heard tales of his adventurous Viking ancestors as well as his family's involvement in the Crusades. Given its association with the Knights Templar, whose headquarters was but a short ride away, it is also likely that he would have been both entertained and educated during his formative years by stories of their heroic exploits and other accomplishments. As a much travelled and cultured nobleman of his day, who also visited the Holy Land, he would have become familiar with the romantic legend of the Grail and been quite capable of comprehending its deeper meaning. Although Henry Sinclair was a leading sea lord of his time, who overcame the challenging geographic and political obstacles involved in asserting, maintaining and expanding his Norwegian earldom, any association with the Templars' cause would definitely have aided and encouraged him in his attempt to reach new lands across the North Atlantic.

Niven Sinclair, the keynote speaker at the Sinclair Symposium in Kirkwall, Orkney, in August 1997, elaborated on possible political and commercial reasons behind the voyage.

There was the urgent need to find new land and new resources and to open up new trade routes for Queen Margrette of Norway whose ships were being harassed in the Baltic by the German merchant traders of Hanseatic League. The speaker also suggested that although Templar wealth may have contributed to the voyage, it was Templar knowledge that was the real treasure that Prince Henry carried across the North Atlantic in 1398. Given that Henry Sinclair's upbringing, travels and associations likely imbued him with the chivalric ideals and aspirations of this brotherhood, it was a legitimate point. Certainly in the references to him in the writings of the Sinclair family historian, one gets the impression that he was a man of exceptional character. There are also hints in the Zeno Narrative in which he is praised by Antonio as "a prince who deserves immortal memory ... for his great bravery and remarkable goodness." These traits are echoed in some of the tales about the benevolent nature of Glooscap. According to Mi'kmaq oral tradition, Glooscap left his island home to come to Nova Scotia where he taught the Mi'kmaq how to better use the resources at hand and he encouraged them to live together in peace.

The Templar interest in harmony is also in evidence in the design used in the building of the Newport Tower on Rhode Island, which early sites researcher James P. Whittall revealed to be based on principles of sacred architecture which the Templars learned in Outremer. It has been suggested more than once that the voyage was in fact the geographic extension of the Knights Templar's hopes of building a new Jerusalem, figuratively and physically, somewhere in the West. The self-same vision a few centuries later resulted in the arrival and spread of Freemasonry in North America. Its ideals, in turn, greatly influenced the thinking of many of the people responsible for drawing up the Constitution of the United States, and in committing themselves to the challenge of creating a new and democratic nation. Unfortunately few people today are aware of the linkage that exists between western democracy and the ideals contained in the legend of the Grail. It has, for too long, remained part of our 'hidden history.'

However, many people would agree that its fundamental themes of individual transformation and of reconnection with the regenerative powers of the universe is as necessary in our time as it was during any era in the past. Emma Jung, author of *The Grail Legend,* has emphasised that the ongoing interest in the Grail legend reflects our need to remain in touch with that which is whole and sacred within ourselves. John Matthews has pointed out in his introduction to *Sources of the Grail* that the story of this chivalric quest, like the tale of the medieval alchemist's preoccupation with trying to transmute base metal into gold, is an allegory for the process whereby we may bring about transformation within ourselves and improve the world in which we live. Both emphasize that the journey involved in the Grail quest is not time specific nor culturally conditioned and is as necessary today as it was during the Middle Ages.

Not surprisingly, interest in and even the search for the physical Grail continues. During the time I was researching and writing this book I came across several references to its alleged discovery or hiding place. Some individuals believe it has already been found in the Middle East or at Glastonbury. Others seem convinced that it still lies buried in a mountainside in southern France, or beneath the roof of Rosslyn Chapel, or in one of several North American sites, including Oak Island. The theory that it was taken across the Atlantic has some people hoping it may be found in Nova Scotia but no one has provded any hard evidence. After digesting some of these claims, I conjured up in my mind a boisterous and colourful debate amongst the various proponents. One could only hope that the wisdom contained in the legend of the Grail would prevail.

11

Excavations and Celebrations

Following their departure from New Ross in 1990, Joan and Ron Harris moved to Waterloo, Ontario. Having run a visitors' hostel on the central Nova Scotia premises for a number of years they had been hoping that the Canadian Hostel Association would take over and continue to manage the property. There was also interest being expressed by some American hostellers in raising funds to purchase it, however, neither of these possibilities materialised and so the house went on the open market. In 1991 the Harrises entered into a business arrangement with Kentville lawyer Walter Newton whereby he had use of the property while it still remained for sale. It was during this period that I first visited the site. By 1994, with no purchaser in sight, Newton managed to rent the house to a local couple, the Pyes. Alva Pye was somewhat familiar with the property and relatives of his had previously rented an apartment in the house for a brief spell during the 1980s. After a short period of renting he managed to arrange for a mortgage and his wife Rose began to operate a private retirement home for seniors on the premises.

Within a year or so, Pye found he was being visited regularly by people who had either heard or read about Joan Harris's unusual discoveries and extraordinary claims. Having found their way to 'the Cross,' Bradley's semi-concealed description of the location of the site, they had landed on the Pye's doorstep drawn by a variety of factors but all sharing a common curiosity to see for themselves what evidence existed of the alleged ruins.

Initially Alva Pye paid little or no attention to such claims since, as far as he was concerned, there was no obvious evidence to support them. By the time he had moved into the house, the rear garden area had become overgrown and very little of anything was noticeable on the ground. After clearing the site of weeds and debris, there was still nothing of consequence to see, certainly nothing that would indicate that a seventeenth-century mansion, let alone a castle or a Viking hall, had once occupied the site. However, a chance meeting with a local resident, retired bank manager Dale Williamson, who was keenly interested in the site's historical and tourist potential, combined with the steady stream of curious visitors and a visit from west coast archaeological psychic George McMullen made Pye sit up and take notice. While his wife continued to run the seniors home, he took to exploring the garden and even carried out an amateur dig on a rectangular plot on higher ground at the south-western end of the lot. Nothing of significance was discovered, but this did not deter Alva Pye's growing interest in the property's possible colourful past.

Following the appearance of a full page article about the site in the local paper, *The Progress Enterprise*, in August 1996, I made another trip to New Ross. The article made extensive references to a possible Sinclair and Templar connection to the site and also to Pye's belief in the existence of a healing well underneath a standing stone in the garden. After an introductory conversation with Pye, I walked around the property once again. Apart from the upright rock, a 3000-pound chunk of granite, and short sections of the lines of stones I had seen a few years earlier, there was not much else on view. He showed me a small collection of artifacts, mostly pieces of stone or metal of indeterminate origin that he had found while digging. From his comments it was obvious that he had become convinced that the standing stone, which I knew from Joan Harris's notes had been found lying in the ground a short distance from where it now stood, marked the location of an ancient holy well. My account of the stone's origin and of the earlier dig by Lloyd Dickie and the Keddys, that had found no evidence whatsoever of an existing or former well, did not seem to deter Pye, who tended to believe that the stone had healing powers. He mentioned that a group of people in Montreal, one of whom claimed she had been miraculously cured of a blood disorder after visiting the site, was interested in developing the property. Not surprisingly this plan came to naught.

Public interest in the site increased following yet another newspaper article about its possible historical significance in the spring of 1997. This article referred to the visit made a few months earlier by George McMullen, whose ability to psychically 'see' evidence of unexcavated historical sites around the world has been verified on numerous occasions. Back in the early 1970s McMullen had become friends with Professor J. Norman Emerson, who taught anthropology at the University of Toronto and was founder of the Ontario Archaeological Society. Introduced to each other by their wives, McMullen had used his psychic ability to help Emerson with a health problem. Impressed by the efficacy of his friend's gift, Emerson then gave McMullen the task of intuitively describing the origin of selected archaeological artifacts and sites, which he did convincingly. During the course of an excavation of what was suspected of having been an Iroquois village in southern Ontario, McMullen was able to throw light on the exact location of a ceremonial long house and the low wooden palisade that surrounded the village. Further excavations proved him right. Brought to public attention by Emerson, who was of the opinion that such intuitive archaeology could be effectively used in conjunction with accepted scientific techniques, McMullen subsequently provided psychically obtained information on a number of archaeological sites around the world. On one occasion his ability helped pinpoint the location of a 1,500-year-old building in the buried Byzantine city of Marea about 70 kilometres north of Alexandria in Egypt. The subject of articles in *Time* and *Maclean's,* as well as in several other magazines, he has also been written about in a number of books, including *One White Crow* which chronicles his work with Professor Emerson. More recently he has helped American archeologists in their efforts to locate the site of the lost colony of Roanoke, Virginia.

Preferring to be known as an intuitive rather than a psychic, McMullen claims that he can mentally project himself up into the air above a given site. From this vantage point he is then able to see it not as it exists in the present but as it was in the past. McMullen admits that his extrasensory ability is not completely accurate, nonetheless he has a reasonably good track record. While visiting the south shore of Nova Scotia with an associate in 1996 in the hope of visiting Oak Island, which he was unable to do, he had driven north to the New Ross site. Without making his identity known to Pye he spent a while on the property tuning in to its past. Although he claims that at the time he consciously knew nothing about a theory connecting Prince Henry Sinclair to the site, his intuitive ability not only enabled him to 'see' an older building on the property but also suggested to him that it had been constructed by Sinclair and his men.

In late January 1997, McMullen sent Pye the following handwritten account of the psychic insights he received:

> The New Ross building which Henry St. Clair lived in during his visit to Nova Scotia was approximately 28 feet wide [8.5 metres] by 30 feet [9 metres] deep. The lower outer walls were built of stone with a wood post interior from floor to ceiling. Above the lower stone walls which were 2 feet [60 cm] thick by 7 feet [2 metres] high, the wood posts formed the upper exterior

walls and provided structural support for the split roofs which were covered with bundles of grass and reeds. The loft was accessed by a staircase constructed from half logs for steps. The sleeping quarters were located in the loft and also below the loft beside the fireplace. Henry St. Clair and other leaders slept in the loft which had a wood floor, men of lower rank and the servants slept on the dirt floor beneath the loft. Across from the fireplace was the dining area furnished with a bench along the opposite wall. There was a dining table with stools. The front of the building (facing the lake) had a door and two openings. The rear of the building (facing the hillside) had just a door. One side of the building had no openings. The opposite side which also had no openings formed the interior wall of the attached stable. There was an opening in the loft by which the interior temperature was controlled. The smaller openings were designed to allow the occupants to defend the building using projectiles. The stone walls offered better protection against attack by hostile natives.

After reading this signed account of his psychic perceptions about the New Ross property and seeing the accompanying hand-drawn sketch of the structure, I had no reason to doubt that McMullen had correctly described a former building on the site, but I was left with a nagging concern about its alleged origin. The building was not at all typical of a fourteenth-century structure and I could not help thinking that what McMullen 'saw' was a version of one of the cabins that I had learned had been built on the property in the 1940s. There was also the distinct possibility that his interpretation of what he had seen had been influenced by something he may have heard earlier about the site. I realised of course that any doubts of mine did not discount the possibility that he had received an authentic independent impression about the origin of the building

Section of archaeological dig underway at New Ross site in 1998.

and in a telephone conversation with him he reaffirmed his conviction that the building was Sinclair related.

Pye now expressed his intention of "properly preserving the property" and only proceeding with any excavations should McMullen approve, however, there was no money to even fence off the site, let alone finance a professional dig. The Nova Scotia Museum of Natural History, whose officials had come to their own opinion about the Harris discoveries many years earlier, was not about to jump in, even if it had the money and staff to spare, which due to government cutbacks it did not. In 1997 Pye began to look for possible partners or even a purchaser. By the end of the year, due to his wife's illness, the seniors home had to be cut back and Pye was in financial trouble. The bank foreclosed and he and his wife moved to Kentville, on the other side of the province, and later to Port Williams. The property remained vacant and neglected and looked in a pretty dilapidated state when I again visited it the following

Archaeological dig on site of stone ‘walls’ at New Ross, 1998.

spring. Although there seemed little likelihood that anyone would be interested in the house as a place to live, there was always the hope that someone, having heard or read about the property's alleged history, might want to see it investigated further. Providence then played a hand and by the time summer 1998 rolled around a new owner had come to the rescue.

Glenn Penoyer is a Toronto businessman who studied under Professor Emerson at the University of Toronto in the 1970s and graduated with a degree in archaeology. In addition, he has also chalked up a PhD in philosophy at the University of Leicester, England. Although involved in managing his family's plumbing business, he was able to spend some time pursuing his primary interest, archaeology. During our first phone conversation I asked him how he had come to know about the New Ross site. He explained that while involved in an archaeological dig in Toronto, which uncovered a thirteenth-century Indian village and burial ground, he had made contact with George McMullen who told him about his earlier visit to Nova Scotia and his perceptions about the New Ross property. Based on this information from McMullen, Penoyer paid a visit and immediately proceeded to purchase it.

By mid-summer, with help from Dale Williamson, he had succeeded in opening up a number of trenches in areas suggested by McMullen. But in spite of taking a painstaking, professional approach and many days of digging trenches in the garden area nothing notable was found to confirm that a centuries-old structure had once stood on the site. When looking over the excavation work I could understand why Penoyer's initial enthusiasm might have been dampened somewhat. Although a number of linear formations of stones were noticeable on the surface no convincing evidence of the foundation walls Joan Harris claimed to have uncovered was apparent in the exposed areas. But in a telephone conversation with him after he had returned to Toronto, Penoyer sounded quite determined to proceed with the dig as soon as his business obligations permitted.

In early October he was able to return to New Ross and set about extending the dig in one specific location, a spot where George McMullen had indicated part of an old wall had already been exposed. Interrupted by heavy rains in its final days, the dig was not as extensive as he had hoped and failed to produce the solid evidence he was looking for. Undaunted by this setback and knowing from training and experience that archaeological work requires patience and persistence, Glen Penoyer planned to proceed with the New Ross excavation project when time and weather allowed. In his favour was the fact that during a meeting with Joan Harris he received first-hand information about her discoveries as well as her collection of artifacts from the site.

Meanwhile, Jack Sinclair made overtures to the company laying a new gas pipeline in north eastern Nova Scotia with the request to be on the lookout for any artifacts or evidence of early stone buildings as they cut and dug their way through the terrain west of Guysborough. He and his brother were involved in celebrations that marked the 600th anniversary of the Sinclair voyage during the summer of 1998. During one of my visits to the Pictou area I had heard that there were two or three communities intending to host some kind of celebrations. There was also talk of a visit to Nova Scotia and New England from a replica ship that was about to be built in Scotland and sailed across the North Atlantic. Plans for the ship never got off the drawing board due to lack of the substantial funding and sufficient lead time. This in no way detracted from other plans in the works in Nova Scotia for commemorating the original voyage.

It was during the final days of these preparations that an unexpected but important discovery in relation to Henry Sinclair's seagoing exploits surfaced. It happened in a very casual and synchronistic way, as these things sometimes do. In spite of several archival searches no reference to any of Sinclair's ships had ever surfaced, although old drawings and museum models of fourteenth-century ships provided a good idea of what they might have looked like. Nonethe-

Drawing of one of Prince Henry Sinclair's ships.

less the absence of any historically accurate description or image of the actual ships used by Sinclair left a serious gap in the research into his alleged voyage to Nova Scotia.

Ron Norquay of Twin City Productions, a promotional products firm in Halifax, is a former navy man and avid collector of maritime memorabilia. On hearing about the planned celebration for the 600th anniversary of the Sinclair voyage he recalled that several years earlier he had come across an etching of a fourteenth-century ship hanging on the wall of an old pub in Edinburgh, Scotland. On inquiring about the origin of the print he was told by the owner of the Heart of Scotland bar in Old Town that it had been in the family's possession for many years and was believed to have been acquired following a clean-out of the archives in Edinburgh Castle. Then on examining it closely Norquay found that writing below the outline of the ship described it as belonging to Earl Henry Sinclair of Orkney. Sensing that it was a rare depiction of a Scottish vessel of the period,

Norquay then struck a bargain with the owner who gave him permission to have it laser copied. On returning home to Halifax he stored it among numerous other items in his collection, and there it rested until Bill Sinclair, who headed up the committee planning celebrations in Nova Scotia, walked into Norquay's office.

On examining a photocopy reproduction of the etching I could see that the ship's design was true to that of northern sea-going vessels in Sinclair's time. It conformed to the findings in a detailed report issued by Robert Green of R.D. Green Engineering, Northfield, New Jersey, who carried out research into the type of vessel most likely used by Sinclair. Although it was not possible to decipher any name on the ship itself, and the writing below the original etching had unfortunately not reproduced, I noted that the ship's stylised bowsprit resembled the sea dragon on Henry's Sinclair's own coat of arms. An even more exciting discovery was the faded and almost obscured outline of the distinctive, eight-pointed Templar Cross on the ship's foresail. Here at last was some solid, visual evidence of the historic connection between Sinclair and the Knights Templar and the association of the Order with his maritime exploits.

Prior to giving a presentation about the Sinclair voyage and the legend of the Grail at Dalhousie University, I learned that two very active organisations had been formed in Nova Scotia to promote greater public awareness of the Sinclair voyage of 1398, not only in the province but also nationally. One of these groups, the Prince Henry Sinclair Society of North America, raised the flag early when its enthusiastic founders D'Elayne and Richard Coleman of New York, who also have property near Guysborough, erected and unveiled a twelve-foot high granite monument at a lookoff site at Half Island Cove in November 1996. Situated on the scenic drive between Guysborough and Canso it overlooks the waters of Chedabucto Bay into which, according to Pohl, the Sinclair expedition had sailed. The inscription on the accompanying plaque provides a general outline of the Sinclair saga. The Colemans were also

Sinclair memorial in Boylston Park, Guysborough, Nova Scotia.

at the forefront of the Prince Henry Project Committee which was set up in 1997 in the United States to encourage and publicise various celebration events south of the border, mostly in New England. Pete Cummings, its secretary, kept the country posted of its plans.

As was to be expected, the Clan Sinclair Association of Canada and the Clan Sinclair Society of Nova Scotia were heavily involved in arranging the 600th anniversary celebrations. In spite of the ongoing debate about the voyage, the Province, through a number of its tourism and culture related agencies, decided to jump on board by way of financial and promotional support, much to the chagrin of some mainstream historians. Organised by the Celebration 600 Committee which was spearheaded by Bill Sinclair and his brother Jack and promoted across Canada by the bagpipe-playing Rory Sinclair of Toronto, the celebrations began appropriately enough with a medieval feast in early June in Halifax. It was given an authentic atmosphere by members of the Society for Creative Anachronism who dressed as

The unveiling of Prince Henry Sinclair Memorial of Boylston Park, Guysborough with Lord Malcolm Sinclair, direct descendant of Prince Henry, Bill Sinclair, President of the Sinclair Clan of Canada and the Norwegian Consul Steiner Engeset.

men and women of the court, served up medieval fare and performed period songs, music and dances. The following month the annual Nova Scotia Tattoo, a major tourist attraction in any year, featured Clan Sinclair and paid a tribute to Prince Henry's voyage. The Halifax Highland games also honoured Clan Sinclair and featured a display of medieval sporting events.

Then during the weekend of July 13-14 the main celebrations were held in the Guysborough area. A second and larger Prince Henry Sinclair Memorial was unveiled and dedicated at the summit of Boylston Park, which is just a short drive outside of the village of Guysborough on route 16. Designed by Scott Sinclair of Halifax in the shape of an upright bow of a fourteenth-century Scottish ship, with a bowsprit replicating the arms of the Earl of Orkney, it received the approval and blessing of the Nova Scotia government which through the Department of Natural Resources and the Department of Education and Culture also financed a good portion of the installation costs. More than four

Chedabucto Bay, Nova Scotia.

metres (13 feet) high, it encases an interpretive exhibit and script which describes the Sinclair saga in English, French, and Mi'kmaq. Several hundred people attended the colourful opening ceremonies at the elevated provincial park site which offers a spectacular, panoramic view of the surrounding countryside that includes the inner reaches of Chedabucto Bay and Guysborough Harbour where Prince Henry is believed to have come ashore. The international gathering was addressed by Lieutenant Governor John Kinley and Deputy Premier Don Downe. Ben Syliboy, Grand Chief of the Mi'kmaq Nation was also represented. Among the special guests was the hereditary chief of Clan Sinclair, the Honourable Malcolm Sinclair, Earl of Caithness, Scotland, a member of the British House of Lords and a direct descendant of Prince Henry Sinclair. And since Henry had been a Norwegian earl at the time of the voyage, the Norwegian government was represented by Consul Steiner J. Engeset. Adding a realistic visual reminder of who and what the occasion was all about was the mail clad, mounted figure of Prince Henry Sinclair himself who rode onto the site accom-

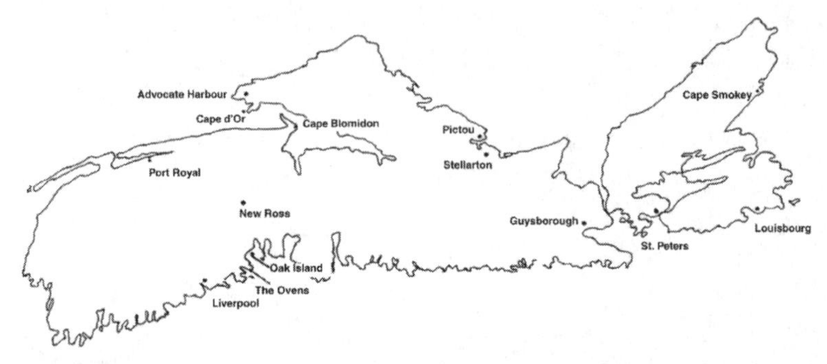

Map of Nova Scotia showing some of the sites associated with the visit of Prince Henry Sinclair

panied by an assorted entourage in medieval garb. It was quite an impressive event and in contrast to the weather which dampened the launching of the Newfoundland Cabot celebrations of a year earlier, it took place in splendid sunshine that sparkled on the blue waters of the bay below. A garden party followed in the village and that night a packed ceilidh, featuring Nova Scotian and Scottish music, dancing and song, was held in the local school. I retired sometime shortly before midnight, exhausted by the day's events while the celebrating, the talk and the tunes continued until the small hours.

Advocate Harbour, which lies snugly below Cape d'Or almost directly due west of Guysborough on the other side of the province, is believed to be the area where Sinclair spent the winter and whence he sailed out into the Bay of Fundy on his way southwest to New England. A local committee headed by Neil St. Clair participated in the 600th anniversary celebrations by way of a concert and an ecumenical church service during the first weekend in August.

The major New England celebrations were held in mid-September in Lincoln, New Hampshire, during the annual Loon Mountain Games, which were also attended by the Earl of Caithness. A feature here was a Sinclair Symposium which presented speakers on a variety of related topics. Westford, Massachusetts, the site of the rock carving of the Gunn knight, also hosted talks on Sinclair and his voyage, the Templars and Masons, Rosslyn Chapel and the Newport Tower. The Westford library put on a special exhibition that included material related to the discovery of the knight carving and the boat stone. Some local schools even participated in an educational project based on the Sinclair saga. As was the case across Canada, Clan Sinclair was the honoured clan at various Highland Games throughout the United States during 1998.

Before leaving the Guysborough area following the July celebrations I made one last visit to Boylston Park to photograph the memorial in the early morning light. Looking out over the inviting vista of tree-covered hills, fertile valleys and waters below, a scene relatively unchanged from Prince Henry's day, it was easy to understand why, if he had climbed that same hill and seen this view of the New World in 1398, he and his associates would have wanted to return to establish a permanent settlement here. His untimely death in the Orkneys a few years later deprived him of that accomplishment.

Of course, the controversy as to whether or not Sinclair ever made such a voyage continues. Perhaps as happened in relation to the once-disputed Viking sagas, this one will also finally be verified by the discovery of some more convincing archaeological evidence, either on the east or west side of the Atlantic. But regardless of what may or may not be discovered in the future, the Sinclair saga, comprised as it is of the Zeno Narrative, the traditional tale of Glooscap, the images in stone of a medieval knight and ship, mysterious ruins in Nova Scotia, the secretive activities of the Templars and the legend of the Grail, has become part of transatlantic lore that connects islands and continents from Scandinavia in the east to New England in the west.

BIBLIOGRAPHY

Alcock, Leslie. *Arthur's Britain.* The Penguin Press, London, 1971.

Allegro, John M.. *The Dead Sea Scrolls.* Penguin Books Ltd.. Middlesex, England, 1956.

Anderson, Peter. "Sinclair Orkney Earldom and Castle." Paper presented at the Sinclair Symposium, Kirkwall, Orkney, August, 1997.

Baigent, Michael and Richard Leigh. *The Temple and the Lodge.* Jonathan Cape, London, 1989.

Barbaro, Marco. *Discendenze Patrizie.* (Museo Correr) Venice, 1536

Bradley, Michael. *Holy Grail Across the Atlantic.* Hounslow Press, Toronto, 1988.

Bucke, Richard Maurice. *Cosmic Consciousness.* University Books Inc., New York, 1961.

Campbell, Joseph. *The Power of Myth.* Anchor Books Doubleday, New York, 1991.

____*The Hero's Journey.* Harper and Row Publishers, San Francisco, 1990.

____*The Masks of God, Creative Mythology.* The Viking Press, New York, 1968.

Cavendish, Richard (Ed). *Man, Myth and Magic: The Illustrated Encyclopaedia of Mythology, Religion and the Unknown.* Marshall Cavendish, Freeport, New York, 1983.

Cayce, Edgar.*The Early Christian Epoch.* The Readings Research Department, Association for Research and Enlightenment, Virginia Beach, Virginia,1976.

Christmas, Peter. *Wejkwapnniaq.* Mi'kmaq Association of Cultural Studies, Sydney, Nova Scotia, 1977.

DeBlois, Albert D. *Micmac Texts*, The Mercury Series, Canadian Museum of Civilisation, Hull, Quebec, 1990.

Degler, Teri. *The Fiery Muse, Creativity and The Spiritual Quest.* Random House, Toronto, 1996.

Delpar, Helen (Ed). *The Discoverers.* McGraw Hill Book Company, New York,1980.

Donovan, Frank R. *The Vikings* American Heritage Publishing Co., Inc., New York, 1964.

Durant, Will. *The Age of Faith, the Story of Civilisation Vol. IV.* Simon and Schuster, New York, 1950.

____*Our Oriental Heritage, the Story of Civilisation Vol. I.* Simon and Schuster New York, 1963

Eliade, Mircea. *Myth and Reality.* Harper and Row, New York, 1968.

Eusebius, of Caesarea. *History of the Church from Christ to Constantine.* Penguin Books Ltd., Middlesex, England, 1981,

Finnan, Mark. *Oak Island Secrets.* Formac Publishing Co. Ltd., Halifax, Nova Scotia, 1995.

____*The First Nova Scotian.* Formac Publishing Co. Ltd., Halifax, Nova Scotia, 1997.

Fitzgerald, Paula. *The Story of the King.* Corinthian Publications, Norfolk, Virginia, 1994.

Forster, Johann Reinhold. *History of the Voyages and Discoveries made in the North.* London, 1786.

Furst, Jeffrey. *Edgar Cayce's Story of Jesus.* Neville Spearman Ltd., London, 1968.

Ganong, W. F. *Crucial Maps.* The University of Toronto Press and the Royal Society of Canada, Toronto, 1964.

Goodrich, Norma Lorre. *King Arthur.* Franklin Watts, New York, 1986.

Halevi, Z'ev ben Shimon. *Kabbalah, Tradition of Hidden Knowledge.* Thames and Hudson, London, 1979.

Hannon, Leslie F. *The Discoverers: The Seafaring Men Who First Touched the Coasts of Canada.* McClelland and Stewart, Toronto, 1971.

Hart, Harriet. *History of the County of Guysborough.* Mika Publishing Co., Belleville, Ontario, 1975.

Hill Elder, Elizabeth. *Joseph of Arimathea.* Real Israel Press, Glastonbury, England, 1996.

Hobbs, William H. *The Fourteenth-Century Discovery of America by Antonio Zeno*. Scientific Monthly vol. 72, January, 1951

Hope, Joan. *A Castle in Nova Scotia.*(Self Published) Kitchener Printing, Kitchner, Ontario, 1997.

Howarth, Stephen W.R. *The Knights Templar*. Atheneum Press, London, 1982.

Humber, Charles J (Ed). *Canada's Native Peoples.* Heirloom Publishing Inc., Mississauga, Ontario, 1988.

Jonas, Hans. *The Gnostic Religion.* Beacon Press, Boston, 1963.

Jost, A.C. Guysborough Sketches and other Essays. Kentville Publishing Co., Kentville, Nova Scotia, 1950.

Jung, Carl Gustav. *Man and His Symbols.* Doubleday and Company Inc., Garden City, New Jersey, 1964.

Jung, Emma and von Franz, Marie Louise. *The Grail Legend.* Sigo Press, Boston, 1986.

Kittler, Glenn D. Egar Cayce on the Dead Sea Scrolls. Coronet Communications Inc., New York, 1970.

Leopold, Caroline. *The History of New Ross.* New Ross Historical Society, New Ross, Nova Scotia, 1966.

Lynch, Michael. *Scotland, A New History.* Pimlico Publishing, London, 1992.

Loomis, R. S. The Grail, From Celtic Myth to Christian Symbolism. University of Wales Press, Cardiff, 1963.

Mackey, Albert, G. *The Encyclopaedia of Freemasonry and its Kindred Sciences.* Moss and Company, Philadelphia, U.S.A., 1875.

Magnusson, Magnus. *The Viking Expansion Westwards*. Bodley Head, London, 1973.

Major, Richard Henry. *The Voyages of the Venetian Brothers, Nicolo and Antonio Zeno, to the Northern Seas, in the XIVth. century*. Hakluyt Society Works, No. 50, London, 1873.

Matthews, John. *Sources of the Grail.* Floris Books, Edinburgh, Scotland 1996.

McKerracher, Archie. *Who Won at Bannockburn. The Highlander.* vol. 32, no. 4, Barrington, Illinois, U.S.A. July 1994.

Murphy, Tim Wallace and Hopkins, Marilyn. *Pilgrimage of Initiation, Friends of Rosslyn*, Edinburgh, 1996.

Ouspensky P. D. *The Psychology of Man's Possible Evolution.* Alfred A. Knopf, New York, 1959.

Outhwaite, Leonard. *Unrolling the Map.* The John Day Company, New York, 1972.

Owen, D. D. R. *The Evolution of the Grail Legend.* Oliver and Boyd, Edinburgh, 1968

Patterson, George D. *A History of the County of Pictou.* Dawson Bros., Montreal, 1877

Phillips, Graham. *The Search for the Grail.* Century Press, London, 1995.

Pohl, Frederick J. *The Sinclair Expedition to Nova Scotia in 1398.* The Pictou Press, Pictou, Nova Scotia, 1950.

____*Prince Henry Sinclair, His Expedition to the New World in 1398.* Clarkson N. Potter Inc., New York, 1974.

Prata, Kathleen R. *Symbols, Guiding Lights along the Journey of Life. Association for Research and Enlightenment*, Virginia Beach,Virginia. 1997.

Rankin. D.J. *A History of the County of Antigonish, Nova Scotia.* The Macmillan Co., Toronto, 1929.

Riley-Smith, Jonathan (Ed). *The Oxford Illustrated History of the Crusades.* Oxford University Press, Oxford, 1995.

Scott, J. *The Early History of Glastonbury.* Boydell Press, London, 1981.

Severin, Tim. "Retracing the First Crusade." National Geographic vol. 176, no. 3, September, 1989.

Sinclair, A. Maclean. *The Sinclairs of Roslin*, Caithness and Goshen. The Examiner Publishing Company, Charlottetown, Prince Edward Island, 1901.

Sinclair, Andrew. *The Sword and the Grail.* Crown Publishers Inc., New York, 1992.

Sinclair, Niven. *Beyond any Shadow of Doub*t. (Self Published), London, 1998.

Smith, J.R. *The Knights of Saint John of Jerusalem and Cyprus.* MacMillan, London, 1987.

Spicer, Stanley P. *Glooscap Legends.* Lancelot Press Ltd., Hantsport, Nova Scotia, 1991.

Steiner, Rudolf. *The Holy Grail.* Extracts from various works compiled by Steven Roboz. Steiner Book Centre, Vancouver, 1984.

TenDam, Hans. *Exploring Reincarnation.* (A. E .J. Wils translator) Arkana, The Penguin Group, London, 1990.

The Scottish Review, Volume XXXII. The Knights Templars in Scotland. Edinburgh, Scotland, 1898.

Weston, Jessie L. *The Quest of the Holy Grail.* Haskell House, New York, U.S.A., 1965.

Whitehead, Ruth Holmes. *Stories From The Six Worlds. Micmac Legends.* Nimbus Publishing, Halifax, Nova Scotia., 1988.

Willoughby, Rupert. *Life in Medieval England (1066-1485).* Pitkin Guides Ltd., Andover, Hampshire, England, 1997.

Wilmshurst W.L. *The Meaning of Masonry.* Gramercy Books, New York, 1995.

Yeats, Frances A. *The Rosicrucian Enlightenment.* Routledge and Kegan Paul, London, 1972.